Computers in Geology— 25 Years of Progress

INTERNATIONAL ASSOCIATION FOR MATHEMATICAL GEOLOGY
STUDIES IN MATHEMATICAL GEOLOGY

1. William B. Size (ed.):
Use and Abuse of Statistical Methods
in the Earth Sciences

2. Lawrence J. Drew:
Oil and Gas Forecasting:
Reflections of a Petroleum Geologist

3. Ricardo A. Olea (ed.):
Geostatistical Glossary and
Multilingual Dictionary

4. Regina L. Hunter and C. John Mann (eds.):
Techniques for Determining Probabilities of
Geologic Events and Processes

5. John C. Davis and Ute Christina Herzfeld (eds.):
Computers in Geology—25 Years of Progress

Computers in Geology—
25 Years of Progress

Edited by

John C. Davis
Kansas Geological Survey

Ute Christina Herzfeld
Scripps Institution of Oceanography

New York   Oxford
OXFORD UNIVERSITY PRESS
1993

Oxford University Press

Oxford   New York   Toronto
Delhi   Bombay   Calcutta   Madras   Karachi
Kuala Lumpur   Singapore   Hong Kong   Tokyo
Nairobi   Dar es Salaam   Cape Town
Melbourne   Auckland   Madrid

and associated companies in
Berlin   Ibadan

Published by Oxford University Press, Inc.,
200 Madison Avenue, New York, New York 10016

Oxford is a registered trademark of Oxford University Press

Library of Congress Cataloging-in-Publication Data
Computers in geology : 25 years of progress
edited by John C. Davis, Ute Christina Herzfeld.
p. cm.—(Studies in mathematical geology ; 5)
Includes bibliographical references and index.
ISBN 0-19-508593-0
1. Geology—Data processing.
2. Geology—Mathematics—Data processing.
I. Herzfeld, Ute Christina.
II. Series.
QE48.8.C63   1993   550'.285—dc20
93-15965

2 4 6 8 9 7 5 3 1

Printed in the United States of America
on acid-free paper

Foreword to the Series

This series of studies in mathematical geology provides contributions from the geomathematical community on topics of special interest in the Earth sciences. As far as possible, each volume in the series will be self-contained and will deal with a specific technique of analysis. For the most part, the results of research will be emphasized. An important part of these studies will be an evaluation of the adequacy and the appropriateness of present geomathematical and geostatistical applications. It is hoped the volumes in this series will become valuable working and research tools in all facets of geology. Each volume will be issued under the auspices of the International Association for Mathematical Geology.

Richard B. McCammon
U.S. Geological Survey
Reston, Virginia, USA

Preface

This book serves multiple functions. Primarily, it demonstrates the current breadth of quantitative applications in the Earth sciences and illustrates how the field of mathematical geology has progressed during the past quarter century. Not coincidentally, it has been 25 years since the International Association for Mathematical Geology was founded. This book also commemorates that beginning in Prague in 1968 and marks the progress of the Association. A society succeeds with the support of its members and thrives on the labors of the few who are most dedicated. The IAMG has benefitted throughout its existence from the unstinting efforts of one such devotee— Daniel F. Merriam. The noted members of the geomathematical community who have contributed to this book are also linked by their association with Dan, to whom the volume is dedicated on the occasion of his 65th birthday and his 25th year of affiliation with the IAMG.

Since the Association's founding, mathematical geology has expanded beyond the application of classical statistical models to geological data and now includes the use of artificial intelligence, computer modeling, simulation, and other procedures not yet developed in 1968. From its relatively obscure roots as a technique for mine evaluation, geostatistics has grown to become a vigorous branch of statistics relevant in all areas of the natural sciences. Computer mapping has evolved from an esoteric task performed at great expense on a limited number of "giant" (at the time) mainframe computers to a routine procedure that runs handily on desk- or lap-top machines. Growth in quantitative procedures has been paralleled by growth in applications and the scientific literature. The *Journal of the International Association for Mathematical Geology* first appeared in 1969 and in 1986 became simply *Mathematical Geology*; from 244 pages in a volume comprising two issues per year it has grown steadily to eight issues per year with well over 1000 pages per volume. In 1975 the IAMG launched *Computers & Geosciences* as a quarterly of some 350 pages per volume. *C&G* has expanded to ten issues per year with 1500 pages per volume. *Nonrenewable Resources* (1992) is the Association's most recent endeavor, a quarterly dedicated to quantitative assessment of the mineral, energy, and resource endowment of the Earth. The IAMG also sponsors the series Studies in Mathematical Geology, established in 1987, in which this volume is the fifth.

Dan Merriam has been instrumental in the IAMG's publishing activities as Editor of both *Mathematical Geology* and *C&G*. Without his tireless efforts, the fledgling journals would not have survived. Dan has served the Association as Secretary-General, President, and Past President. He has organized innumerable conferences and symposia (including the Silver Anniversary Meeting), usually editing and arranging the published proceedings. Dan's myriad activities on behalf of the IAMG are affectionately recounted by my co-editor. Those fortunate enough to know Dan will recognize him in her words:

> "Having just returned from the International Geological Congress in Kyoto, I am thinking of the latest experience of Dan's everlasting engagement for the sake of 'geomathematics' (naming our discipline by the term he prefers over 'mathematical geology,' which is more than a subtlety in meaning). Dan Merriam has shaped geomathematics. He is always organizing new meetings, spreading out work, always has a few essays to review, a manuscript to finish, or a question ready for you. Dan's office is a place to find new papers, discover the latest book in the discipline, learn about upcoming events. Dan has a fascination with geomathematics and the ability to draw people around him into it—to work day and night on several projects. A typical scene at a dinner table: [Dan] 'Is this a working dinner?' Next, he opens his bag and hands some work to everyone around the table, something to read...to write...to review...to plan. With Dan, these things are fun. But with all the importance of geomathematics, Dan is and will always remain a field geologist. At any of my visits to Kansas, a field trip in the geology has been essential. With this *Festschrift*, we are thanking Dan for his involvement in mathematical geology and, in particular, for the many things he has taught us."

Ute Christina Herzfeld
La Jolla, California, September 1992

This book, then, carries a triple dedication: To the science of mathematical geology for the progress it has made over the past 25 years, to the Silver Anniversary of the International Association for Mathematical Geology, and to Daniel F. Merriam, who has given so much to both.

John C. Davis
Lawrence, Kansas, January 1993

Acknowledgments

The Editors would like to express their appreciation to the authors and to those who contributed to the preparation of this volume. Foremost among the latter is Jo Anne De Graffenreid of the Kansas Geological Survey. Jo Anne served as copy editor, designer, and typesetter for the book and did much to improve grammatical aspects of many of the papers. To prepare the camera-ready copy, she mastered the intricacies of the computer typesetting languages TEX and LATEX as well as the arcana of electronic file transfer via the international computer network. Without her determination and selfless efforts, the book would never have seen the light of day.

The Editors would also like to thank several other staff members of the Kansas Geological Survey, particularly Renate Hensiek, who made numerous camera shots and photo-reductions and drafted or redrafted illustrations from distant authors who found themselves bereft of such services. Survey Librarian Janice Sorensen tracked down missing citations and completed references where these were not provided by the authors. Early in the project, Geoff Bohling converted diskettes from IBM to Macintosh format, translating files from a bewildering array of word processing programs into something more-or-less readable.

Drs. John Doveton and Ricardo Olea of the Kansas Geological Survey served as reviewers, as did Helmut Mayer of Scripps Institution of Oceanography. To them and to our anonymous reviewers we extend our heartfelt thanks. We also thank Series Editor Dr. Richard B. McCammon, not least for saving us from ourselves and providing the title, *Computers in Geology—*

25 Years of Progress, in lieu of our zestier favorite: "The Digital Pick—25 Years of Progress Using Computers in Geology."

For the color plates which enhance this book we gratefully acknowledge financial support from the Geological Survey of Canada and from ARCO Exploration and Production Technology, Plano, Texas, USA. Finally, we thank the Kansas Geological Survey and its Director, Dr. Lee Gerhard, for providing the logistical, computer, and personnel resources necessary to produce this volume.

JCD
UCH

Contents

Contributors

- F.P. Agterberg
  Geological Survey of Canada
  Ottawa K1A0E8, CANADA

- G.C. Bohling
  Kansas Geological Survey
  The University of Kansas
  Lawrence, KS 66047, USA

- G.F. Bonham-Carter
  Geological Survey of Canada
  Ottawa K1A0E8, CANADA

- H. A. F. Chaves
  Department of Geology/Geophysics
  University of the State of Rio de Janeiro
  CEP 20550 Rio de Janeiro, BRAZIL

- Q. Cheng
  Geological Survey of Canada
  Ottawa K1A0E8, CANADA

- D.R. Collins
  Kansas Geological Survey
  The University of Kansas
  Lawrence, KS 66047, USA

- R.G. Craig
  Department of Geology
  Kent State University
  Kent, Ohio 44242, USA

- J.M. Cubitt
  Geochem Group Ltd.
  Chester CH4 8RD, UK

- J.C. Davis
  Kansas Geological Survey
  The University of Kansas
  Lawrence, KS 66047, USA

- J.H. Doveton
  Kansas Geological Survey
  The University of Kansas
  Lawrence, KS 66047, USA

- A. Flexer
  Tel Aviv University
  Tel Aviv, ISRAEL

- D. Gill
  Geological Survey of Israel
  Jerusalem, ISRAEL

- John W. Harbaugh
  Department of Applied Earth Sciences
  Stanford University
  Stanford, California 94305, USA

- J.E. Harff
  Institute for Baltic Sea Research
  O–2530 Warnemünde, FRG

- Ute Christina Herzfeld
  Scripps Institution of Oceanography
  University of California at San Diego
  La Jolla, California 92093, USA

- S. Heynisch
  Institute of Geology
  Geophysics and Geoinformatics
  Free University of Berlin
  D-1000 Berlin 46, FRG

- Michael Edward Hohn
  West Virginia Geological and Economic Survey
  Morgantown, WV 26507, USA

- Marek Kacewicz
  ARCO Exploration and Production Technology
  Plano, Texas 75075, USA

- S.B. Kelly
  Geochem Group Ltd.
  Aberdeen AB1 4LF, UK

- M. Levinger
  A.D.N. Computer Systems Ltd.
  Jerusalem, ISRAEL

- C. John Mann
  Department of Geology
  University of Illinois
  Urbana, IL 61801, USA

- Helmut Mayer
  Scripps Institution of Oceanography
  University of California at San Diego
  La Jolla, CA 92093, USA

- Richard B. McCammon
  U.S. Geological Survey
  National Center 920
  Reston, Virginia 22092, USA

- Donald B. McIntyre
  Honorary Fellow
  Univs. St. Andrews & Edinburgh
  Kinfauns, Perth PH2 7LD
  Scotland, UK

- K.G. McKenzie
  Department of Geology
  University of Melbourne
  Parkville, Vic., AUSTRALIA

- Václav Němec
  Consultant
  K rybníčkům 17
  100 00 Praha 10, CZECH REPUBLIC

- R.A. Olea
  Kansas Geological Survey
  The University of Kansas
  Lawrence, KS 66047, USA

- Richard A. Reyment
  Paleontologiska Institutionen
  Uppsala Universitet, SWEDEN &
  Département des Sciences de l'Evolution
  USTL, Montpellier, FRANCE

- Joseph E. Robinson
  Department of Geology
  Syracuse University
  Syracuse, New York 13244, USA

- V. Rohrlich
  Centre de Géologie Générale et Minière
  Ecole Nationale des Mines de Paris
  77350 Fontainebleau, FRANCE &
  Technion, Haifa, ISRAEL

- Henri Sanguinetti
  GEOVAL
  Sydney, NSW 2000, AUSTRALIA

- W. Schwarzacher
  School of Geosciences
  The Queen's University of Belfast
  Belfast BT7 1NN, NORTHERN IRELAND

- W. Skala
  Institute of Geology
  Geophysics and Geoinformatics
  Free University of Berlin
  D-1000 Berlin 46, FRG

- H. Teil
  Centre de Géologie Générale et Minière
  Ecole Nationale des Mines de Paris
  77350 Fontainebleau, FRANCE

- A. Toister
  Tel Aviv University
  Tel Aviv, ISRAEL

- J. Tourenq
  Laboratoire de Géologie des Bassins Sédimentaires
  Université P. et M. Curie
  75005 Paris, FRANCE

- B. Vardi
  A.D.N. Computer Systems Ltd.
  Jerusalem, ISRAEL

- Hans Wackernagel
  Centre de Géostatistique
  Ecole des Mines de Paris
  77305 Fontainebleau, FRANCE

- Johannes Wendebourg
  Department of Applied Earth Sciences
  Stanford University
  Stanford, California 94305, USA

- E. H. Timothy Whitten
  Earth Resources Centre
  University of Exeter
  Exeter EX4 4QE, UK

- D.F. Wright
  Geological Survey of Canada
  Ottawa K1A0E8, CANADA

	1968–1972	1972–1976	1976–1980	1980–1984
President	A.B. Vistelius	R.A. Reyment	D.F. Merriam	E.H.T. Whitten
Past President	W.C. Krumbein served as "Past President"	A.B. Vistelius	R.A. Reyment	D.F. Merriam
Vice President	G.S. Watson	A.T. Bharu-cha-Reid	G. Hill	D.G. Hawkins
Treasurers				
Eastern	V. Němec	V. Němec	V. Němec	V.T. Vuchev
Western	T.V. Loudon	J.C. Davis	J.C. Davis	R.B. McCammon
Secretary General	R.A. Reyment	D.F. Merriam	E.H.T. Whitten	J.C. Davis
Council Members	F.P. Agterberg	H.A.F. Chaves	F.P. Agterberg	A.C. Cook
	D.G. Krige	A.C. Cook[1]	K.L. Burns	I. Djafarov[1]
	G. Matheron	J.E. Klovan	D. Gill	R.J. Howarth
	S.C. Robinson[1]	P. Laffite	D.M. Hawkins	N. Nishiwaki
	D.A. Rodionov	G. Lea	R.J. Howarth	S.N. Sadooni
	S.P. Sengupta	D. Marsal	G. de Marsily[1]	A.B. Vistelius
Editors-in-Chief	E.H.T. Whitten	E.H.T. Whitten	W. Schwarzacher	
Math. Geology	D.F. Merriam	D.F. Merriam	R.B. McCammon	T.A. Jones
Computers & Geosciences	—	D.F. Merriam	D.F. Merriam	J.M. Cubitt D.F. Merriam
News Letter	G. Lea	J.C. Davis	J.C. Davis	J.C. Davis

	1984–1989	1989–1992	1992–1996
President	J.C. Davis	R.B. McCammon	M.E. Hohn
Past President	E.H.T. Whitten	J.C. Davis	R.B. McCammon
Vice President	P. Switzer	J. Aitchison	C.-J.F. Chung
Treasurers			
Eastern	V. Němec	V. Němec	V. Němec
Western	M.E. Hohn	J.O. Kork	J.O. Kork
Secretary General	R.B. McCammon	M.E. Hohn	R.A. Olea
Council Members	I. Clark	P.I. Brooker	M. Alfaro
	J.M. Cubitt	C.-J.F. Chung	E. Grunsky
	D. Gill	J.M. Cubitt	V. Pawlowsky
	A. Marechal	K.H. Esbensen	J.C. Tipper
	H.M. Parker[1]	A. Marechal	G. Verly
	R. Sinding-Larsen	N. Nishiwaki[1]	P.D. Zhao[1]
Editors-in-Chief	G. Williams	R.A. Olea	D. Zhou
Math. Geology	C.J. Mann	R. Ehrlich	R. Ehrlich
Computers & Geosciences	D.F. Merriam	D.F. Merriam	D.F. Merriam
News Letter	J.C. Davis	J.R. Carr	J.R. Carr
Studies in Mathematical Geology	R.B. McCammon	R.B. McCammon	R.B. McCammon
Nonrenewable Resources	—	—	R.B. McCammon

[1]IGC Special Councillor

Computers in Geology—
25 Years of Progress

1

INTRODUCTION

Václav Němec

Computers in Geology—25 Years of Progress

Friends and associates of Daniel F. Merriam have prepared this volume in Dan's honor to commemorate his 65*th* birthday and mark the 25*th* anniversary of the International Association for Mathematical Geology. This compendium is in the tradition of the *Festschriften* issued by European universities and scholarly organizations to honor an individual who has bequeathed an exceptional legacy to his students, associates, and his discipline. Certainly Dan has made such an impact on geology, and particularly mathematical geology. It is a great privilege for me to write the introduction to this *Festschrift*. The editors are to be congratulated for their idea to collect and to publish so many representative scientific articles written by famous authors of several generations.

Dan Merriam is the most famous mathematical geologist in the world. This statement will probably provoke some criticism against an over-glorification of Dan. Some readers will have their own candidates (including themselves) for such a top position. I would like to bring a testimony that the statement is correct and far from an *ad hoc* judgment only for this solemn occasion.

To the First Contacts East–West

It may be of interest to describe how I became acquainted with Dan. In my opinion this will show how thin and delicate was the original tissue of invisible ties which helped to build up the first contacts among Western and Eastern colleagues in the completely new discipline of mathematical geology.

The role of Dan Merriam in opening and increasing these contacts has been very active indeed.

In the Fall 1964 I was on a family visit in the United States. This was—after the coup of Prague in 1948—my first travel to the free Western world. With some experience in computerized evaluation of ore deposits, I was curious to see the application of computers in geology and to meet colleagues who had experience with introducing statistical methods into regular estimation of ore reserves. I had very useful contacts in Colorado and in Arizona. In Tucson I visited the real birthplace of the APCOM symposia. The University of Arizona was just preparing the fifth symposium of this series for 1965 and they honored me by an invitation to take an active part in it. I have been always very grateful to Professors W.C. Lacy and W.C. Peters from Tucson and to R.F. Hewlett (then working at Golden) who were among the organizers of the first APCOM symposia and who not only arranged the invitation for me but who gave me various examples and inspiration of how to use computers in solving geological and mining problems. I got also the address of a Russian colleague, Ivan P. Sharapoff, whom I met later in 1965 at Sochi and who really opened for me the way to many other scientists in his own country, and also to a few colleagues in other countries (as France or South Africa).

Finally I did not appear at the APCOM '65 but three papers from Czechoslovakia were published in the Symposium proceedings. In the list of APCOM '65 participants you will not find the name of D.F. Merriam; however, two authors of this *Festschrift* were present: John W. Harbaugh and Frits P. Agterberg. At that time John already had regular working contacts with Dan and probably he drew the attention to computerized colleagues from Czechoslovakia when in 1966 both Dan and John were preparing their first visit to Eastern Europe. They wrote letters to me and to another Czech author, Blahomil Soukup. But it was only my colleague who received the letter—our addresses were uncomplete and the letter for me did not reach me at all. We knew that our visitors would arrive to Prague one definite Saturday or Sunday in October 1966 from Krakow. I wrote a letter to Krakow asking some local colleagues there to transmit our message with some useful information (correct addresses, telephone numbers). No answer appeared and we had no other option than to be waiting for both days at the Main railway station of Prague paying special attention to the arrivals of all possible trains which could be used for traveling from Krakow to Prague. The expected visitors did not arrive.

My First Meeting with Dan Merriam

About three days later I had in my office a phone call from my father who told me that a telegram just arrived at my home from Salzburg announcing the arrival of D.F. Merriam in Prague for the same day. My office was then located about 20 km outside Prague. The telephone liaison there was constantly overloaded and in the daytime it was almost a miracle that I got the call and was able to return in time to Prague. Since early afternoon I was with my colleague awaiting the arrival of our visitor. The working hours were already over when a call came—somebody on the other side passed the mouth-piece to Dr. Merriam who announced he had just come and was awaiting us at the entrance. We thought that it was the entrance of our enterprise and we quickly ran there to meet him. But there was not any evidence of him. The only chance was that another call would give us some additional information. After about a quarter of an hour came a new call. It was my father announcing that Professor Merriam just called to my home and that he was waiting near the entrance of the Hotel Europe at about 4 km distance from us, just in the center of Prague at the famous Wenceslaus Square. In about 15 minutes we arrived there. Blahomil knew from some journal a photograph of Professor Merriam and he was sure he would recognize him. Some people were standing nearby the entrance, one typical American with a full beard among them, others in the hall of the hotel, but evidently nobody was Dr. Merriam. He was not registered in the hotel as a guest and apparently he did not use the telephone at the hotel reception. I had to call my father again asking if perhaps some further sign of life from Professor Merriam came. Fortunately after another 20 minutes when I repeated my question my father acknowledged that Dr. Merriam was really awaiting us just in front of the Hotel Europe. Evidently it was that American gentleman with the full beard—absolutely different from the person known from the picture. Unexpectedly we had discovered that Dan sometimes changes his face! After this discovery I never had problems recognizing him although some intervals between our meetings were relatively long.

But we are still with Dan on his first brief visit to Prague. It is already the evening; we take something to eat in a restaurant nearby and Dan is explaining why he came so late to Prague that day. During the previous week he was with Professor Harbaugh in Poland. They had the opportunity to travel directly from Krakow to Austria and during the transit through Czechoslovakia they had no problems with any passport control. In Salzburg Dan decided to make a one-day trip with a rented car to Prague. But on the border the Czech passport control discovered the difference between his real face and his photograph in his passport and refused to let him enter the territory of Czechoslovakia unless he shaved his beard or had an additional photograph in his travel documents. Dan decided for solution No. 2 and returned to the nearest Austrian town to arrange his new picture. Several hours were simply lost.

It was too late for Dan to go back to Salzburg the same night. We tried in vain to arrange a hotel room for him in Prague, but it was possible to book a room in a motel about 50 km from Prague on the way to Austria. We went there with our cars. We spent a wonderful evening. I still remember a delicious dinner in a local restaurant followed by a very long discussion on problems of mutual interest. Certainly we spoke also about the International Geological Congress 1968 which was already in preparation in Czechoslovakia. And we were sure in the moment of our farewell—already a couple of hours after midnight—that in less than two years we should meet again in Prague.

International Geological Congress 1968— Foundation of the IAMG

How many beginning changes in public life, how many great expectations during the famous Prague Spring of 1968. I was completely absorbed by my own work (computerized models of large deposits of industrial minerals were not at all an easy job at that time) as well as with preparations for the International Geological Congress. I served as member of a committee for "other problems" where for the first time in the history of the IGC mathematical geology constituted a considerable part. Professor R.A. Reyment—evidently following the advice of Dan—asked me to serve on an international committee to prepare the foundation of an international association for application of mathematical methods and computers in geology (the definite name had to be chosen during the foundation meeting) and as the only Czechoslovak member of this committee, I also had some duties as quartermaster for the committee members as well as for the committee sessions in coordination with the program committee of the Congress. I also served as manager of an 11-day pre-Congress excursion (another of the authors from this *Festschrift*, E.H.T. Whitten was among the participants). It was just at the opening ceremony the 19*th* of August 1968 where I met Dan Merriam again and where I met for the first time many other members of the preparatory committee. We were about eight who took lunch together that day, incidentally four future presidents of the IAMG among us: A.B. Vistelius, R.A. Reyment, D.F. Merriam, and R.B. McCammon. We then spent a lovely afternoon visiting the castle of Prague (I still remember splendid sunshine and unusual visibility when I was showing our visitors the beautiful panorama of Prague from the castle area), we tasted some famous Czech beer in a small pub nearby and in the evening I accompanied Dan and John Harbaugh to their hotel. The menu list was only in Czech and Dan decided to order for everybody another meal with a random selection and to make the real choice later *in natura*.

The next day was the first real working day of the Congress. I met Dan and other colleagues many times, mostly in corridors of the building of Technical University where the Congress was held. Dan's extrovert ability

Outside the Hotel Europe, Prague, August 21, 1968

—Photo by A.J. "Bert" Rowell

to attract people and to bring them together was evident. In the afternoon Blahomil Soukup and I invited Dan with John Harbaugh to visit the head-quarters of Geoindustria to show them some results of our work. We started to think how to organize some special program for the next days, especially for the weekend. What happy expectations, what happy days. . . .

In the next night I woke up very, very early because of some very unusual and heavy uproar of airplanes constantly repeated in almost regular waves. In the darkness of the night I observed from the window long series of low-flying planes. The radio Prague was announcing repeatedly only the entry of armies of five countries of the Warsaw Treaty—without any approval of legal Czechoslovak authorities. I spent several hours of that day with Professor Reyment—his hotel was not far from my home. I remember also my own analysis of the situation I presumed at that very sad morning, "This is the end of communism in the world!" The history verified it in more than 21 years. For that day nobody was able to forecast what would happen in the next hours and days. The municipal transport was completely out of order, all main squares and crossroads were full of Soviet tanks. Happy days were over. . . .

The day after the invasion—22*nd* of August 1968—special buses brought Congress guests to the Technical University. This was the only day when the

Congress tried to continue normal work according to the official program. Fortunately one meeting of our preparatory committee was scheduled for that morning. The original idea was to prepare everything for the foundation meeting scheduled for another day. The program also reserved space and time for a closing meeting of the new Council. We had to decide and we chose to use the first scheduled meeting directly to found the International Association for Mathematical Geology. About 20 persons present elected the first IAMG Council. Already this voting made it clear that all present members were able to differentiate between the political interest of Brezhnev on one side and the desire for democracy of the common people as well as the needs of science on the other side. Andrey Borisovich Vistelius was unanimously elected as first President of the IAMG despite the fact that soldiers of his native country were just taking part in a brutal occupation of Czechoslovak territory. Other elected officers: Vice Presidents: W.C. Krumbein, G.S. Watson, Secretary General: R.A. Reyment, Treasurers: T.V. Loudon, V. Němec. Ordinary members of the Council: E.H.T. Whitten, D.A. Rodionov, D.G. Krige, G. Matheron, F.P. Agterberg, S.P. Sengupta. Dan Merriam was elected as first Editor-in-Chief.

The new IAMG Council had no time to arrange any additional meeting during the Congress which had to be prematurely closed the next day. No time and also no desire to drink some champagne, no occasion and no desire to continue in talks concerning the IAMG. The security of Congress participants had to be assured and the only possible way was to organize transports of special buses and convoys of cars to Vienna and some cities in West Germany. (The airport of Prague was closed for public transport for almost one month.) At noon for the last time I shook hands with Dan and many other new colleagues. In the afternoon A.B. Vistelius, E.H.T. Whitten, and myself represented the IAMG at the premature closing meeting of the Council of the International Union of Geological Sciences where the foundation of our new Association was announced and its affiliation to the IUGS unanimously approved.

D.F. Merriam and the IAMG

It may seem inappropriate to write in this Introduction so much about the IAMG and so little about Dan Merriam. In reality there are so many intersections of Dan and of the IAMG. Dan is one of the "top-fidelity" members of the IAMG. He has constantly served in the IAMG Council since its beginning. He served also as Secretary General (1972–1976) and as President of the IAMG (1976–1980)—in both functions immediately replacing R.A. Reyment (the real convener and founder of the Association). Dan started two highly successful journals, *Mathematical Geology* and *Computers & Geosciences* (in the last one he still continues as Editor-in-Chief). To bring these journals to life—that was probably the most tedious job among the duties of any member of all the IAMG Councils. The journals seem nowadays to be the best children of the IAMG and we cannot forget that it was Dan who served not only as a godfather but as a real father and a real mother and a real accoucheur in one person!

Dan has been a tireless organizer and convener of numerous international meetings—starting with the famous series of colloquia at Lawrence, Kansas, and continuing later with the Geochautauquas at Syracuse University, New York, and elsewhere. Dan has been a spirit and a living personification of the IAMG. Writing about the IAMG means that this is at the same time writing about Dan.

From Other Remembrances

Numerous remembrances of my contacts with Dan after the tragic Congress of Prague can be added. Thanks to Dan I got an invitation from the Kansas Geological Survey to work in the department of mineral resources in the academic year 1969/70. For almost 10 months I had the pleasure to work in the same building as Dan, being in close contact with him in all the IAMG business of that time. I appreciated very much his friendly help during the first days of my acclimatization in Lawrence as well as his help with editing some of my papers and articles prepared for publication. I admired his systematic work in organizing colloquia (I took part in two of them). In his tremendous work he never refused any help—and not only to me but also to dozens of other people in his environment. Many other authors in this *Festschrift* were working during that happy year in Lawrence: Frits Agterberg, Hernani Chaves, John Davis, John Doveton, Joe Robinson. Many others came for a visit or took part in the colloquia: John Cubitt, John Harbaugh, John Mann, Dick McCammon, Tim Whitten.

I remember also our two visits to Wilson ("the Czech capital of Kansas") as well as other interesting field trips and excursions with Dan. How many unforgettable parties Dan organized in his house with the so-efficient help of his wife Annie. I took part also in a flight on a very small plane (for four persons only) to the south of Kansas. At that time I was very intensively engaged in my own research on regular structural patterns and Dan recognized very well that the observations from such a flight may be very inspiring for me. During our return flight to Lawrence Dan served as pilot. In fact my life was completely in Dan's hands, but I felt myself very safe.

Shortly after my return to Prague from Kansas Dan revisited Czechoslovakia to take part in the geomathematical section of the Mining Příbram Symposium. When organizing these regular international meetings (annually in the period 1968–1973, after 1973 every second year) I have had always Dan as a very good model for my own organizational work.

Since 1971 some international congresses (sedimentological at Heidelberg, 1971, and Nice, 1975; geological in Paris, 1980, Moscow, 1984, and Washington, D.C., 1989) were practically the only chance for me to meet Dan again. The most pleasant meeting took place at Nice: the *Palais des Congrès* was just across the street from the seashore and Dan was spending there every free minute. He found there also the inspiration for the cover of the proceedings volume edited later by him—a naked pretty girl. Evidently as Secretary General he was thinking of the benefit to the Association. In accordance with the Statutes, the domicile of the IAMG is where the Secretary General is conducting his business. It was therefore my duty to follow him!

In 1986—thanks to the support of the IAMG—I had again the opportunity to revisit America. I spent a couple of days in Wichita, Kansas, in Dan's home. His style of work was not changed at all. In Fall 1991 Dan revisited Prague and Příbram. We met at Wenceslaus Square very near the Hotel Europe but without any problem of identification. I had the impression that, under the protection of the good Czech patron St. Wenceslaus sitting on his horse in the famous statue nearby, old happy days had returned again to my beloved Prague. One circle of life has been closed but another circle has started. And Dan continues to work in the pilot's cabin. . . .

Conclusions

In case the above-mentioned personal remembrances appear as insufficient support for the statement that Dan Merriam really is the most famous mathematical geologist in the world, some other reasons can be added:

- In spite of occasional changes of his face, Dan never changed his personal devotion to developing and promoting mathematical geology all over the world.

- He has been always, without any personal profit, helping his colleagues of many generations and anywhere in the world to introduce mathematics and computers into their own work.

- He has discovered and prepared for further work, to the benefit of the IAMG and mathematical geology, many followers from the whole world.

- He has never known any specific borders or frontiers to scientific research and the development of mathematical geology.

- He has been able to overcome the most tedious problems in his work as Editor-in-Chief of the official journals of the IAMG.

- He has always remained a very good geologist who never left his own original field work and he has always tried to use computers and mathematical methods only as very efficient means to obtain purely geological results.

- He has been a prolific contributor to the geological literature and was co-author, with John Harbaugh, of the seminal volume *Computer Applications in Stratigraphic Analysis*, published in 1968. Dan's bibliography includes over 180 articles, of which 80 have been written on topics in mathematical geology. He also has edited over 20 books and symposium volumes on different aspects of mathematical geology.

Briefly—the almost *incredible impact* of Daniel F. Merriam on many generations of geologists and mathematical geologists is not any *ad hoc* vision but an *historical fact*.

Acknowledgments

It is my great pleasure to express sincerest thanks to Daniel F. Merriam for all his work, help, and impact realized by him for the benefit of mathematical geology. This is not only my personal acknowledgment, it is not only the acknowledgment of all authors and editors of this *Festschrift*, it is really the acknowledgment of all generations of mathematical geologists of the world. *Ad multos annos*, Dan!

For myself, another acknowledgment should be added to the editors of this *Festschrift*. I am deeply honored by their idea of choosing me as the "very first author" in this volume and also for their technical editorial work with my contribution. After all, their efforts show how fruitful were the seeds cultivated by Dan Merriam.

Prague, August 1992

2

WEIGHTS OF EVIDENCE MODELING AND WEIGHTED LOGISTIC REGRESSION FOR MINERAL POTENTIAL MAPPING*

F.P. Agterberg, G.F. Bonham-Carter,
Q. Cheng and D.F. Wright

During the past few years, we have developed a method of weights of evidence modeling for mineral potential mapping (*cf.* Agterberg, 1989; Bonham-Carter *et al.*, 1990). In this paper, weights of evidence modeling and logistic regression are applied to occurrences of hydrothermal vents on the ocean floor, East Pacific Rise near 21° N. For comparison, logistic regression is also applied to occurrences of gold deposits in Meguma Terrane, Nova Scotia.

The volcanic, tectonic, and hydrothermal processes along the central axis of the East Pacific Rise at 21° N were originally studied by Ballard *et al.* (1981). Their maps were previously taken as the starting point for a pilot project on estimation of the probability of occurrence of polymetallic massive sulfide deposits on the ocean floor (Agterberg and Franklin, 1987). In the earlier work, presence or absence of deposits in relatively large square cells was related to explanatory variables quantified for small square cells (pixels) by means of stepwise multiple regression and logistic regression. In this paper, weights of evidence modeling and weighted logistic regression are applied to the same maps but a geographic information system (Intera–TYDAC SPANS, 1991) was used to create polygons for combinations of maps. These polygons can be classified taking the different classes from each map. Probabilities estimated for the resulting "unique conditions" can be classified and displayed. The vents are correlated with only a few patterns and it is relatively easy to interpret the weights and final probability maps

---

*Geological Survey of Canada Contribution No. 45291.

in terms of the underlying volcanic, tectonic, and hydrothermal processes. The vents are situated along the central axis of the rise together with the youngest volcanics. They occur at approximately the same depth below sea level, tend to be associated with pillow flows rather than sheet flows, and with absence of fissures which are more prominent in older volcanics.

Contrary to weights of evidence modeling, weighted logistic regression (*cf.* Agterberg, 1992, for discussion of algorithm) can be applied when the explanatory variables are not conditionally independent. This method was previously applied by Reddy *et al.* (1991) to volcanogenic massive sulfide deposits in the Snow Lake area of Manitoba. The assumptions underlying these methods will be evaluated in detail for the seafloor example.

The gold deposits in Meguma Terrane, Nova Scotia, were previously used for weights of evidence modeling (Bonham-Carter *et al.*, 1988; Wright, 1988; Agterberg *et al.*, 1990; Bonham-Carter *et al.*, 1990). It will be shown in this paper that similar results are obtained when weighted logistic regression is used. The degree of fit of the different statistical models is evaluated for all applications in this paper. A difference between the two examples of application is that the number of hydrothermal vents on the seafloor is small in comparison to the number of Nova Scotia gold deposits. For this reason, the posterior probabilities have greater relative precision in the Nova Scotia example.

Hydrothermal Vents on the Ocean Floor

The maps used for this example are shown in Plate 1 (color illustrations are grouped together; Plates 2–5 follow Plate 1). The volcanics on the East Pacific Rise are of two types: (1) pillow flows, and (2) sheet flows. There are three age classes. Ballard *et al.* (1981) determined relative age by measuring relative amounts of sediments deposited on top of the volcanics. The resulting six litho-age units are shown in Plate 1a together with the occurrences of 13 hydrothermal vents. The other map patterns of Plate 1 are for topography (depth to sea bottom, Pl. 1b), distance to contact between youngest pillow flows and youngest sheet flows (Pl. 1c), and distance to fissures (Pl. 1d). The first step in weights of evidence modeling consisted of constructing binary patterns which are relatively strongly correlated with the vents. Five binary patterns for the seafloor example are shown in Plate 2. Each pattern has positive weight W^+ for presence of a feature and negative weight W^- for its absence. The contrast $C = W^+ - W^-$ is a measure of the strength of correlation between the vents and a binary pattern. A binary pattern can be constructed by maximizing the contrast. For example, from Plate 1b for topography it can be seen that nearly all (12 of 13) vents belong to a single 20-m topographic interval and this prompted the choice of the binary pattern of Plate 2b. A table of contrast versus corridor width can be

Plates

Plate 1: Patterns used for example of occurrence of hydrothermal vents on the seafloor (East Pacific Rise, 21° N; based on Fig. 5 of Ballard *et al.*, 1981). (a) litho-age units and hydrothermal vents (dots); relative age classes are 1.0–1.4 (youngest), 1.4–1.7 (intermediate), and 1.7–2.0 (oldest); (b) topography (depth below sea level); (c) corridors around contact between youngest pillow and sheet flows; (d) corridors around fissures.

Plate 2: Binary patterns derived from Plate 1 (study area as in Pl. 1a). Weights for presence or absence of features were calculated for 0.01 km^2 unit cell size. (a) age, $W^+ = 2.251$ for presence of youngest volcanics, $W^- = -1.231$; (b) topography, $W^+ = 2.037$ for presence within zone with water depth between 2580 m and 2600 m, $W^- = -0.811$; (c) rock type, $W^+ = 0.632$ for presence of pillow flows, $W^- = -0.338$ for presence of sheet flows; (d) proximity to contact between youngest volcanics, $W^+ = 2.259$ for points within 20 m, $W^- = -0.570$; (e) absence/presence of fissures, $W^+ = 0.178$ for points at distances greater than 110 m, $W^- = -0.097$ for points within 110 m.

Plate 3: Seafloor example. (a) Posterior probability map with eight unique conditions for the overlap of first three binary patterns of Plate 2, unit cell size $= 0.01$ km^2; (b) t–value map corresponding to Plate 3a (t–value is ratio of posterior probability and its standard deviation); (c) posterior probability map with 31 unique conditions for the five binary patterns of Plate 2, unit cell size $= 0.001$ km^2.

Plate 4: Weighted logistic regression applied to seafloor example. (a) Posterior probability map with 31 unique conditions for the five binary patterns of Plate 2, unit cell size $= 0.001$ km^2 (*cf.* Pl. 3c); (b) t–value map corresponding to Plate 4a; (c) Posterior probability map with 196 unique conditions for modified logistic model. See text and caption of Table 3 for further explanation.

Plate 5: Weighted logistic regression applied to gold deposits (circles) in Meguma Terrane, Nova Scotia. (a) Posterior probability map with 91 unique conditions for seven binary patterns without missing data, unit cell size $=$ 1 km^2; (b) t–value map corresponding to Plate 5a.

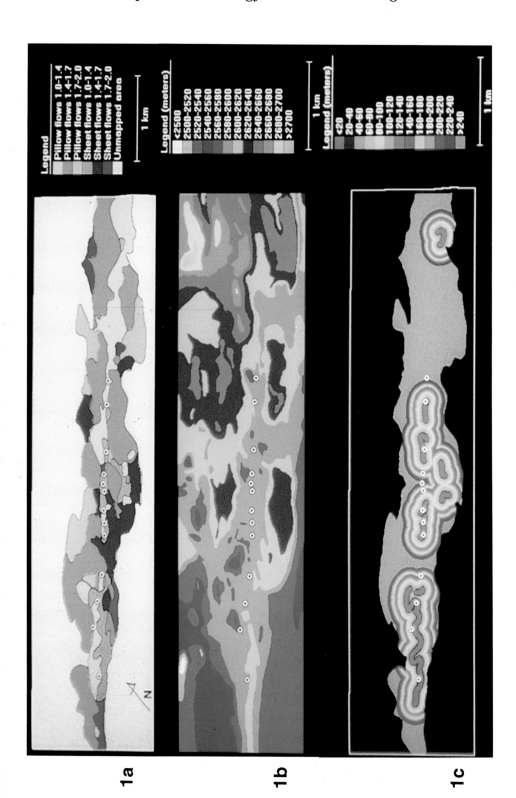

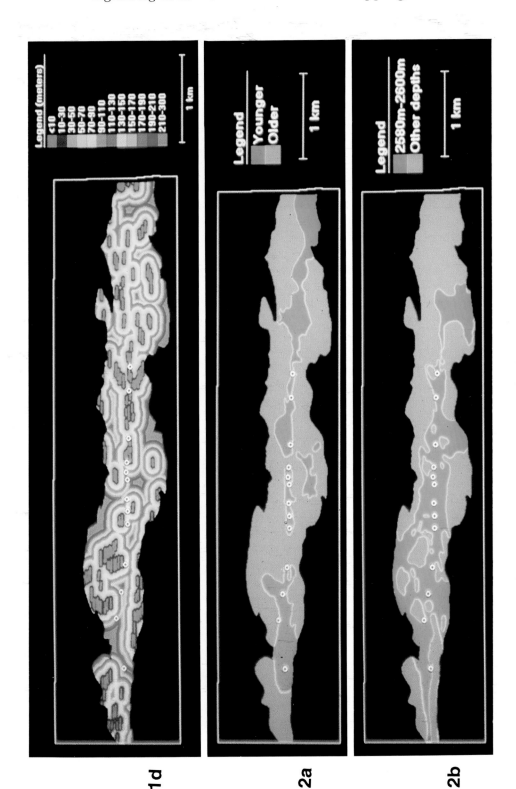

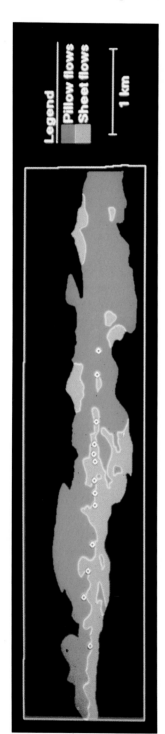

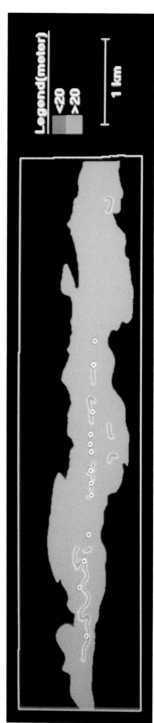

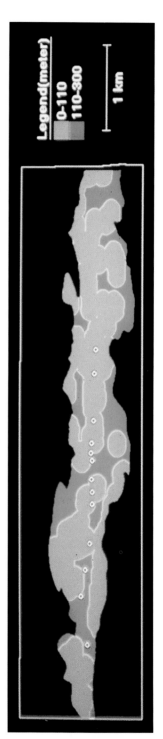

2c

2d

2e

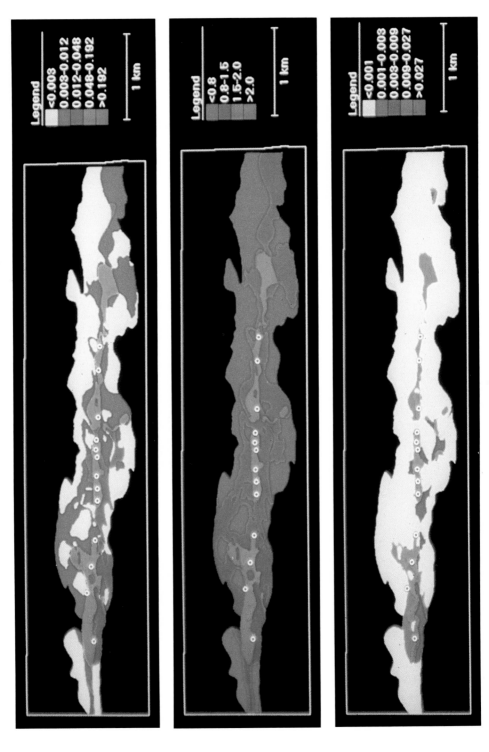

3a

3b

3c

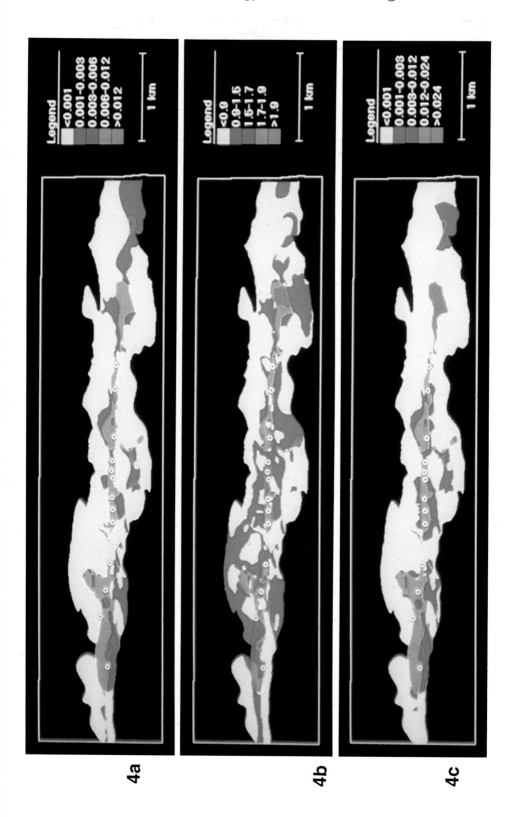

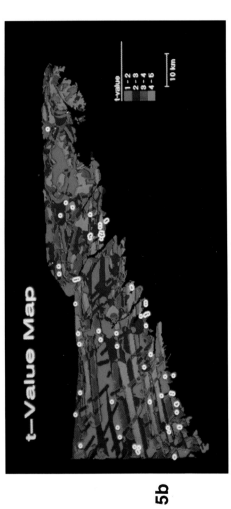

Table 1: Positive Weights (W^+), Contrasts (C), and Standard
Deviations (s) for Corridors around Contact between Youngest Volcanics
(Area measured in km^2. Last column shows t–value of
$C[t = C/s(C)]$. The 13 corridors are displayed in Plate 1c.)

No.	Width	Area	Vents	W^+	$s(W^+)$	C	$s(C)$	$C/s(C)$
1	20m	0.197	6	2.259	0.416	2.829	0.561	5.041
2	40m	0.338	7	1.863	0.382	2.549	0.559	4.557
3	60m	0.532	8	1.538	0.356	2.352	0.572	4.112
4	80m	0.721	8	1.232	0.356	1.989	0.572	3.481
5	100m	0.931	8	0.973	0.355	1.664	0.571	2.912
6	120m	1.097	8	0.808	0.355	1.443	0.571	2.526
7	140m	1.281	9	0.771	0.335	1.564	0.602	2.598
8	160m	1.480	11	0.827	0.303	2.237	0.769	2.908
9	180m	1.660	11	0.712	0.303	2.047	0.769	2.661
10	200m	1.808	11	0.626	0.302	1.896	0.769	2.464
11	220m	2.000	11	0.525	0.302	1.702	0.769	2.212
12	240m	2.164	12	0.533	0.290	2.317	1.041	2.225
13	> 240m	3.984	13					

used for deciding on the binary pattern for distance from a linear feature.
For example, in Table 1 the contact between youngest pillows and sheets
(Pl. 1c) has the largest contrast for corridor no. 1 which was selected for the
binary pattern of Plate 2d. The fissure binary pattern (Pl. 2e) is also for the
corridor with the largest contrast. Weights and contrasts for the binary patterns of Plate 2 are summarized in Table 2. From the standard deviations
it can be seen that the correlation between fissures and vents is probably
not significant. The binary pattern of Plate 2e is only weakly correlated
with the vents. In weights of evidence modeling, the binary patterns are

Table 2: Weights and Contrasts (with Standard Deviations)
for Five Binary Patterns of Plate 2

Pattern	W^+	$s(W^+)$	W^-	$s(W^-)$	C	$s(C)$	$C/s(C)$
Age	2.251	1.000	−1.231	0.290	3.481	1.041	3.343
Topography	2.037	1.000	−0.811	0.290	2.848	1.041	2.735
Contact	2.259	0.415	−0.570	0.378	2.829	0.561	5.041
Rock type	0.632	0.410	−0.338	0.378	0.970	0.558	1.740
Fissures	0.178	0.448	−0.097	0.354	0.275	0.571	0.481

combined with one another by the addition of weights for very small unit cells where the features are either present or absent. From a statistical point of view, this addition is only permitted if the binary patterns are conditionally independent of the vent pattern. Chi-squared statistics for conditional independence testing (Agterberg, 1992) cannot be used here, because the required frequencies of points are too small. It is likely that the binary patterns are not conditionally independent. For example, the age (Pl. 2a), rock type (Pl. 2c) and contact corridor (Pl. 2d) patterns were constructed from the litho-age units of Plate 1a. This assumption is corroborated by performing the following pattern correlation analysis. Yule's measure of association Q for binary variables resembles the product–moment correlation coefficient in that it is equal to zero if there is no correlation and cannot exceed one (for exact linear relationship) in absolute value. It is close to zero for the three pairs: age–topography (0.19), age–rock type (0.18), and topography–rock type (-0.02). In absolute value it is relatively large for contact corridor correlated with age (0.95), topography (0.53), and rock type (0.53), respectively. These results suggest that the contact corridor pattern may be redundant, as will be demonstrated later by statistical tests.

Posterior Probability Maps

Plate 3a is a posterior probability map for a 0.01-km^2 unit cell based on only three binary patterns (age, topography, and rock type). The prior probability in this application was set equal to 0.033 for number of vents ($= 13$) divided by total area ($= 398.4$ unit cells). The binary patterns of Plate 2 also can be regarded as posterior probability maps. Presence of single features gives posterior probabilities of 0.111 (age), 0.073 (topography), and 0.061 (rock type), respectively. These probabilities are for presence of a vent within a 0.01-km^2 subarea at a particular place where an indicator pattern is present. In general, combining p binary patterns gives 2^p possible combinations for the unique conditions. Plate 3a is based on eight unique conditions with probabilities equal to 0.000, 0.001, 0.006, 0.011, 0.015, 0.030, 0.171, and 0.360. The uncertainties of these probabilities are relatively large, as shown in the corresponding t–value map of Plate 3b where every posterior probability was divided by its standard deviation.

Plate 3c is the posterior probability map for a 0.001-km^2 unit cell using all five binary patterns of Plate 2. Although the patterns of Plates 3a and 3c are similar, a more detailed analysis shows that the results of these two applications of the weights of evidence method are different. Plate 3c is based on 31 unique conditions (one of the possible 32 combinations of five features is not represented), with probabilities ranging from 0.000 to 0.352. The unit cell for Plate 3c is ten times as small as the one used for Plate 3a. Because the posterior probabilities cover approximately the same range of values, this means that the probability of finding a vent per 0.01-km^2 unit

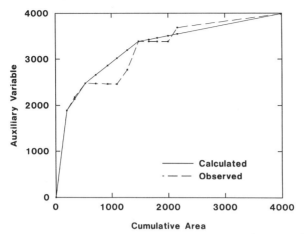

Figure 1: Seafloor example, analysis of relationship between hydrothermal vents and contact between youngest volcanics (*cf.* Table 1). Auxiliary variable $y = A \cdot \exp(W^+)$ is plotted against cumulative area A measured in units of 0.001 km^2. The first derivative dy_c/dA of fitted curve y_c provides estimates of four values of variable weight $W^+(A)$ that depends on distance from the contact. See text for further explanation.

cell in the unique conditions with the largest posterior probabilities is about ten times greater in the situation of Plate 3c. We will show later (see sections on weighted logistic regression and goodness-of-fit test) that the model of Plate 3a provides a good fit, whereas the model underlying Plate 3c overestimates the posterior probabilities in the most favorable areas because of lack of conditional independence of the contact corridor binary pattern.

Analysis of Contact Corridor Pattern

The contrast in Table 1 has secondary maxima for corridors 8 and 12. Although the positive weights for these other corridors are less than that for the first corridor used for Plate 2d, their areas are larger. Expected number of vents within a corridor is equal to the product of corridor area and posterior probability. For this reason, a wider corridor (*e.g.*, no. 8) can also be selected as a binary pattern. Another method of modeling the relationship between vents and contact is to estimate weights for the intersections of successive corridors ("classes") shown in Plate 1c.

Figure 1 was derived from the data of Table 1 for classes of contact corridors as follows. An auxiliary variable $y = A \cdot \exp(W^+)$ is plotted against cumulative area A. Agterberg and Bonham-Carter (1990) have shown that the natural logarithm of the first derivative dy_c/dA of a curve y_c fitted to y may provide a good estimate of $W^+(A)$ representing a variable weight that depends on distance from the contact. Suppose m distinct weights are calculated for m classes of distance instead of the two weights corresponding to the two classes of a binary pattern. The observed values of Table 1 (and Fig. 1) are for increasingly wide corridors. Adjoining classes with the smallest difference in y can be combined repeatedly until only m new classes are retained. The result of this iterative process for $m = 4$ is shown in Figure 1 as four straight-line segments approximating y_c. The slopes of the four straight lines can be used to estimate the following four weights: 2.259 (for class 1, as before), 0.566 (for classes 2 and 3), -0.043 (for classes 4 to 8), and -1.431 (for remainder of study area). This pattern suggests an approximately linear decrease in weight with distance from the contact. This, in turn, implies that the probability of finding a vent within a small cell would decrease exponentially with distance. It will be shown next how these results can be incorporated in the modeling.

Weighted Logistic Regression

Weights of evidence modeling and logistic regression with the observations weighted according to their areas of the corresponding unique conditions are different types of application of the loglinear model (*cf.* Andersen, 1990). In weighted logistic regression, the patterns are not necessarily conditionally independent as in weights of evidence modeling. Plate 4a shows posterior probabilities for a 0.001 km² unit cell using the same five binary patterns of Plate 3c. The probabilities of Plate 4a range from 0.000 to 0.054. For the most favorable unique conditions, they are nearly ten times as small as the corresponding values that resulted from applying the weights of evidence method to the five binary patterns. In this respect, the posterior probabilities resulting from weighted logistic regression are close to those obtained by applying the weights of evidence method to three binary patterns only (*cf.* Pl. 3a). These results indicate that the large probabilities that arose when the weights of evidence method was used with the five binary variables are, indeed, too large because of lack of conditional independence. The logistic regression coefficients and their standard deviations are shown in Table 3. The t–value map for Plate 4a is shown in Plate 4b.

Weighted logistic regression can also be used in situations where the explanatory variables have many classes or are continuous. In the discussion of Figure 1, it was suggested that probability of occurrence of vents decreases exponentially with distance from contact. In order to incorporate this exponential decrease in the logistic model, a new explanatory variable was created by assigning values decreasing from 13 to 1 to the 13 classes used for Figure 1 (*cf.* Pl. 1c). Combining this new ordinal variable with the previous four binary variables resulted in an increase in the number of unique conditions (from 31 to 196). Plate 4c shows the posterior probability map for this new model. In general, the pattern of Plate 4c is close to the one of Plate 4a. Although the relationship between vents and contact was

Table 3: Regression Coefficients for Logistic Model (B) and
Modified Logistic Model (B') with Standard Deviations

[The value of x in B' for contact between youngest volcanics ranges
from 13 (corridor no. 1) to 1 (corridor no. 13).]

Pattern	B	$s(B)$	B'	$s(B')$
Age	2.862	1.076	2.979	1.086
Topography	2.388	1.050	2.458	1.051
Contact	1.114	0.604	$0.145x$	0.579
Rock type	0.350	0.584	0.420	0.591
Fissures	0.139	0.579	0.062	0.076

modeled in more detail, the overall effect of this refinement becomes small
when it is combined with the relationships of the vents with age, elevation,
and rock type (*cf.* Table 3).

Goodness-of-Fit Test

The degree of fit of several models is evaluated in Figure 2. The posterior
probability is plotted in the horizontal direction. The product of posterior
probability and area per unique condition provides theoretical values for
frequency of vents. Corresponding observed frequencies can be obtained
by counting the number of vents per unique condition. Theoretical and
observed frequencies were converted to relative frequencies by dividing by
total number of vents ($= 13$). If a model is good, the predicted total number
of vents should be close to 13. This condition is nearly satisfied in Figures 2a
(weights of evidence modeling using three binary patterns) and 2b (weighted
logistic regression using five binary patterns). In the situation of Figure 2a,
the model predicts 14.0 vents which is one too many; the model of Figure 2b
predicts 12.6 vents—slightly less than 13. The Kolmogorov–Smirnov (K–S)
test can be used to evaluate the largest difference between observed and
expected relative frequencies. In Figure 2a, the absolute value of the largest
difference is 0.081. In a two-tailed test with eight observations, this value
should not exceed 0.454 with a probability of 95%. The corresponding 95%
confidence level for Figure 2b with 31 observations is 0.238 which also is
greater than the observed value of 0.099 in this diagram. It may be concluded
that the models tested in Figures 2a and 2b provide a good fit.

On the other hand, the degree of fit of the models underlying Figures 2c
and 2d is poor. Figure 2c corresponds to Plate 3c for which it was already
shown that the five binary patterns are not conditionally independent. The
predicted total number of vents is 37.6, which is nearly three times too large.

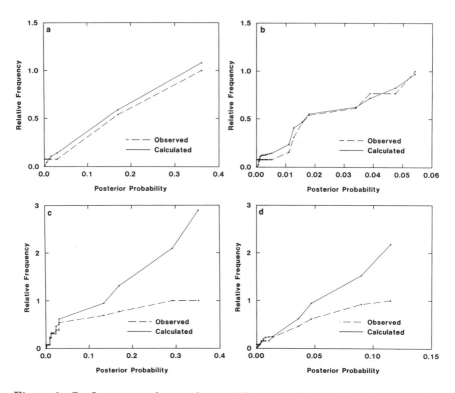

Figure 2: Seafloor example, goodness-of-fit tests. Observed and estimated relative frequencies versus posterior probabilities from (a) Plate 3a, (b) Plate 4a, (c) Plate 3c, and (d) pattern similar to Plate 3c obtained after using contact corridor no. 8 instead of no. 1 for the contact binary pattern.

Moreover, the absolute value of the largest difference (= 1.892) in Figure 2c exceeds the 95% confidence level (= 0.238) in the K–S test. Figure 2d is for a probability map (not shown) derived from five binary patterns in which the contact pattern was for the wider corridor comprising classes 1 through 8 in Plate 1c. The expected total number of vents then is 28.4, which is more than twice the observed total (= 13). The absolute value of the largest difference (= 1.184) in Figure 2d exceeds the 95% confidence level (= 0.238) for a good fit.

The largest posterior probabilities in Figures 2c and 2d are 0.352 and 0.115, respectively. Differences between observed and calculated frequencies do not exceed the 95% confidence level of the K–S test except for the three or four largest posterior probabilities. The models underlying Figures 2c and 2d provide a good fit except in the most favorable unique conditions where the frequencies of vents are overestimated by a wide margin.

The preceding application of the K–S test differs from other applications of this test because in our application the model also predicts total number of discrete events. Normally a non-zero difference between observed and expected frequencies at the largest value does not arise because the observations originate from an infinitely large population. In a strict sense, the Kolmogorov–Smirnov test statistics may only be used when the total number of discrete events is correctly predicted. The approximate K–S test used in this paper loses its validity when the expected relative frequency is not approximately equal to 1.0 at the largest value. Note that in Bonham-Carter *et al.* (1990) the K–S test was applied, but the theoretical as well as the observed cumulative frequencies were constrained to reach a maximum of 1.0. This had the advantage of satisfying the assumptions for the K–S test, but the disadvantage of failing to recognize theoretical frequencies that are too large.

Also note that possible undiscovered deposits are not considered in the goodness-of-fit test. The reason that results of weights of evidence modeling and logistic regression are useful for mineral potential mapping is that the estimated weights are approximately independent of undiscovered deposits in a study region provided that the known deposits can be regarded as a random subset of all (known + unknown) deposits in the region. Only the prior probability in weights of evidence modeling and the constant term in logistic regression depend strongly on undiscovered deposits (*cf.* Agterberg, 1992).

Gold Deposits in Central Nova Scotia

In the weighted logistic regression, 68 gold deposits were related to the following seven binary patterns (*cf.* Bonham-Carter *et al.*, 1990): (1) proximity to anticlinal axes, (2) Au in balsam fir, (3) contact between Goldenville and Halifax Formations, (4) Goldenville Formation, (5) Devonian granite contact zone, (6) lake sediment signature, and (7) NW lineaments. The assumption of conditional independence is slightly violated in this application. For example, weights of evidence modeling for a 1-km^2 unit cell on these seven binary patterns results in a predicted total number of deposits equal to 75.2, which exceeds the observed total by nearly 10%. It is noted that patterns (2) and (6) are missing in parts of the area. In weights of evidence modeling, the weights can be estimated for patterns with missing data by omitting areas that are unknown from the weight calculations. In logistic regression, this procedure would result in significant loss of information because coefficients for all patterns are estimated simultaneously; thus, omitting areas with missing data would eliminate these regions from estimation entirely. For this reason patterns (2) and (6) were modified so that, in regions where the patterns are missing, they were treated as being

Table 4: Weights and Contrasts (with Standard Deviations) for Seven
Binary Patterns Related to Gold Deposits in Meguma Terrane, Nova Scotia

[Regression coefficients for logistic model (B) and their standard deviations
are shown in last two columns. First row (pattern no. 0) is for constant
term in weighted logistic regression.]

Pattern No.	W^+	$s(W^+)$	W^-	$s(W^-)$	C	$s(C)$	B	$s(B)$
0							−6.172	0.501
1	0.563	0.143	−0.829	0.244	1.392	0.283	1.260	0.301
2	0.836	0.210	−0.293	0.160	1.129	0.264	1.322	0.267
3	0.367	0.174	−0.268	0.173	0.635	0.246	0.288	0.266
4	0.311	0.128	−1.474	0.448	1.784	0.466	1.290	0.505
5	0.223	0.306	−0.038	0.134	0.261	0.334	0.505	0.343
6	1.423	0.343	−0.375	0.259	1.798	0.430	0.652	0.383
7	0.041	0.271	−0.010	0.138	0.051	0.304	0.015	0.309

"not present." Logistic regression on the resulting revised data set predicts
64.3 gold deposits—slightly less than 68.

The weights, their standard deviations, and contrasts of the weights of
evidence modeling are compared to the estimated logistic regression coefficients in Table 4. Plate 5a shows the logistic posterior probability map
which is similar to weights of evidence modeling results previously shown
in Bonham-Carter *et al.* (1990). Plate 5b shows the posterior probabilities
divided by their standard deviations (t–value map). A significant difference
between Plate 5b and the t–value maps for the seafloor example (Pls. 3b and
4b) is that the values in Plate 5b are relatively large. In an approximate
signifance test based on the normal distribution in standard form, a t–value
greater than 1.645 indicates that the corresponding posterior probability is
greater than 0 with a probability of 95%. This greater degree of precision is
due to the larger number of occurrences for the Nova Scotia example.

Finally, Figure 3 is for evaluation of the goodness of fit of the logistic
model of Plate 5. The absolute value of the largest difference between expected and observed relative frequencies is 0.0775. This is less than the
Kolmogorov–Smirnov statistic (= 0.1426; 95% two-tailed test) and it may
be concluded that the fit of the logistic model is good.

Concluding Remarks

Care should be taken in weights of evidence modeling to avoid bias caused by
predictive patterns that are mutually interrelated, because violations of the
conditional independence assumption usually lead to overestimation of the

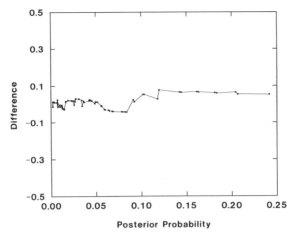

Figure 3: Goodness-of-fit test applied to logistic model for gold deposits, Meguma Terrane (Pl. 5a). The difference between observed and theoretical relative frequencies is plotted against posterior probability. See text for further explanation.

largest posterior probabilities. The problem of bias is avoided when weighted logistic regression is used. In general, the drawbacks of regression are that it cannot be applied without making assumptions about missing values unless all explanatory patterns are fully known for a study area. Moreover, the standard deviations of regression coefficients can be unreasonably large if there is multicollinearity. The latter problems are of minor significance in this paper where the logistic model produced satisfactory results in all applications.

It is suggested in this paper that both weights of evidence and logistic regression solutions be routinely compared. The weights of evidence method yields readily interpreted positive and negative weights and is a straightforward method for determining optimal cutoffs for the creation of binary patterns and for handling missing data. On the other hand, logistic regression provides a check on the effects of lack of conditional independence, in addition to the χ^2- and K–S tests suggested for the weights of evidence method.

References

Agterberg, F.P., 1989, Computer programs for mineral exploration: *Science*, v. 245, p. 76–81.

Agterberg, F.P., 1992, Combining indicator patterns in weights of evidence modeling for resource evaluation: *Nonrenewable Resources*, v. 1, no. 1, p. 35–50.

Agterberg, F.P. and Bonham-Carter, G.F., 1990, Deriving weights of evidence from geoscience contour maps for the prediction of discrete events, *in* TUB-Dokumentation Kongresse und Tagungen No. 51, *Proceedings 22nd APCOM Symposium*, Berlin, September 1990: Tech. Univ. Berlin, v. 2, p. 381–396.

Agterberg, F.P., Bonham-Carter, G.F. and Wright, D.F., 1990, Statistical pattern integration for mineral exploration, in Gaál, G. and Merriam, D.F., (eds.), *Computer Applications in Resource Exploration, Prediction and Assessment for Metals and Petroleum:* Pergamon Press, Oxford, p. 1–21.

Agterberg, F.P. and Franklin, J.M., 1987, Estimation of the probability of occurrence of polymetallic massive sulfide deposits on the ocean floor, in Teleki, P.G. *et al.*, (eds.), *Marine Minerals:* Reidel, Dordrecht, p. 467–483.

Andersen, E.B., 1990, *The Statistical Analysis of Categorical Data:* Springer–Verlag, Berlin, 523 pp.

Ballard, R.D., Francheteau, J., Juteau, T., Rangan, C. and Norwark, W., 1981, East Pacific Rise at 21° N: The oceanic, tectonic, and hydrothermal processes of the central axis: *Earth and Planetary Sci. Letters*, v. 55, p. 1–10.

Bonham-Carter, G.F., Agterberg, F.P. and Wright, D.F., 1988, Integration of geological datasets for gold exploration in Nova Scotia: *Photogrammetry and Remote Sensing*, v. 54, no. 11, p. 1585–1592.

Bonham-Carter, G.F., Agterberg, F.P. and Wright, D.F., 1990, Weights of evidence modelling: A new approach to mapping mineral potential, in Agterberg, F.P. and Bonham-Carter, G.F., (eds.), *Statistical Applications in the Earth Sciences:* Geol. Survey of Canada, Paper 89–9, p. 171–183.

Reddy, R.K.T., Agterberg, F.P. and Bonham-Carter, G.F., 1991, *Application of GIS-based Logistic Models to Base-metal Potential Mapping in Snow Lake Area, Manitoba:* Proceedings, 3rd Canadian Conference on GIS, Ottawa, Canada, March 18–21, 1991, p. 607–619.

Intera–TYDAC, 1991, *SPANS Users Guide, Version 5.2:* Intera–TYDAC Technologies Inc., Ottawa, Canada, 4 volumes.

Wright, D.F., 1988, *Data Integration and Geochemical Evaluation of Meguma Terrane, Nova Scotia, for Gold Mineralization:* Unpublished M.Sc. thesis, University of Ottawa, 82 pp.

3

GOLD PROSPECTING WITH FACTORIAL COKRIGING IN THE LIMOUSIN, FRANCE

Hans Wackernagel
and Henri Sanguinetti

In geochemical prospecting for gold a major difficulty is that many values are below the chemical detection limit. Tracers for gold thus play an important role in the evaluation of multivariate geochemical data. In this case study we apply geostatistical methods presented in Wackernagel (1988) to multielement exploration data from a prospect near Limoges, France. The analysis relies upon a metallogenetic model by Bonnemaison and Marcoux (1987, 1990) describing auriferous mineralization in shear zones of the Limousin.

Concepts of "Anomaly"

The aim of geochemical exploration is to find deposits of raw materials. What is a deposit? It is a *geological anomaly* which has a significant average content of a given raw material and enough spatial extension to have economic value. The geological body defined by an anomaly is generally buried at a specific depth and may be detectable at the surface through indices. These indices, which we shall call *superficial anomalies*, are disposed in three manners: at isolated locations, along faults, and as dispersion halos.

These two definitions of the word "anomaly" correspond to a vision of the geological phenomenon in its full continuity. Yet in exploration geochemistry only a discrete perception of the phenomenon is possible through samples taken along a regularly meshed grid. A superficial anomaly thus can be apprehended by one or several samples or it can escape the grip of the geochemist when it is located between the nodes of the mesh.

Table 1: Anomaly Concepts

3D	2D	Samples
Geological Anomaly $\longmapsto$	Superficial Anomaly $\longmapsto$	Geochemical Anomaly
	Isolated spot $\rightarrow$	Point
Deposit $\rightarrow$	Fault	
	Aureola $\rightarrow$	Group of points

A geochemical anomaly, in the strict sense, only exists at the nodes of the sampling grid and we shall distinguish between:

a *pointwise anomaly* defined on a single sample, and
a *groupwise anomaly* defined on several neighboring samples.

This distinction is important both upstream, for the geological interpretation of geochemical measurements, and downstream, at the level of geostatistical manipulation of the data. It will condition an exploration strategy on the basis of the data representations used in this case study.

A pointwise anomaly, *i.e.*, a high, isolated value of the material being sought, will correspond either to a geological phenomenon of limited extent or to a well hidden deposit. It is therefore always important to examine the highest values, even if they are isolated, but it is likely that intense sampling around pointwise anomalies will prove futile.

A different situation arises when two or more values are above the average, without any of them being necessarily an extreme value. With a grouped anomaly, the associated geological phenomenon is likely to have a significant spatial extension and possibly the samples are within a dispersion aureola of an important deposit.

Table 1 summarizes the different concepts and shows the passage from one type of anomaly to another, when moving from a three-dimensional (underground) vision of the phenomenon to its two-dimensional perception at the surface and, more specifically, to the discrete mesh of sampling points.

The geochemical exploration strategy derived from this conceptualization aims at concentrating efforts on pointwise and groupwise anomalies. Pointwise anomalies can be spotted on maps of circles whose sizes are proportional to values, while groupwise anomalies are best shown by geostatistical filtering operations.

Auriferous Shear Zones

Bonnemaison and Marcoux (1987, 1990) propose a three-stage metallogenetic model which is used for gold prospecting in the Hercynian basement
of France. The three stages are linked with different mineralogical occurrences of gold: either gold is invisible to the naked eye (early stage) or it is
fine-grained (intermediate stage) or it occurs as nuggets (late stage).

In the early stage, gold is concentrated in the form of auriferous sulfides
within lenses along shear zones, which are typically about 10 m wide and
several kilometers long at the surface and have a potential vertical extension
of several hundreds of meters. In the intermediate stage the faults of the
shear zone are filled with different vein minerals which subsequently are
crushed during deformation and fine-grained gold is found in these facies.
The late stage takes place within structures created by the two previous
stages and leads to the formation of gold nuggets that occasionally have a
size of several millimeters.

This metallogenetic model synthesizes a historical vision of different gold
deposits located in shear zones around the world. It is interesting for its geometric content with respect to what can be called the "morphology of the
gold values" or, more precisely, the morphology of the regionalized variable. It shows that interpretation of the data should not emphasize only the
highest gold grades. This would amount to focusing interest only on those
parts of the regionalized variable explained by the late stage of formation
in which nuggets are found. Efforts should also concentrate on describing
areas of lower values stemming from fine-grained or invisible gold, as economic concentrations also form during the early and intermediate stages.

Maps of two types are used in this case study:

1. A map of circles whose areas are proportional to the measured values.
   This simple graphical representation easily identifies the highest gold
   values. It also gives a first idea of the spatial organization of the values.
   When this spatial organization is complex, maps produced by kriging
   will be a desirable complement.

2. A map produced by kriging the short-range component. This map
   amplifies aggregations at a local scale, suppressing large-scale regional
   variation as well as isolated high values. It should facilitate identification of groupwise anomalies.

Kriging the long-range component is also useful for suppressing small-
scale variation and emphasizing large-sized structures which can often be
interpreted by comparison with a geological map. Such cartographical representations have been prepared for gold and its pathfinder elements, either
treating them individually or combining them, but not all are reported here.

Table 2: Linear Correlation Coefficients

	Au	As	Pb	Zn	Ni	Co	Cr	Sb	Li
As	0.16								
Pb	0.09	0.21							
Zn	0.02	−0.02	0.15						
Ni	0.09	−0.04	0.00	0.61					
Co	0.02	−0.12	−0.14	0.72	0.83				
Cr	0.06	−0.07	−0.04	0.58	0.92	0.85			
Sb	0.27	0.30	0.12	0.42	0.44	0.42	0.42		
Li	−0.12	0.11	0.29	0.08	−0.10	−0.14	−0.15	−0.18	
Be	−0.07	0.13	0.31	−0.05	−0.27	−0.24	−0.29	−0.25	0.69

The Charriéras Data Set

The French mining company COGEMA, which owns the Le Bourneix gold mine located south of the city of Limoges, has carried out several geochemical prospecting campaigns in the Limousin. In a prospect known as "Charriéras," located within the region defined by the 1:50,000 geological map "Châteauneuf-la-Forêt" described by Chenevoy (1983), 2541 samples were collected following a nearly regular grid with a mesh size of about 150 m. The samples were analyzed for 12 elements, among which Bi had no value and Ag had only two values above the detection limit, so that they were discarded from subsequent studies.

Gold (Au) has 67% of the analyses below the detection limit and Sb has 34%. The other eight elements (As, Pb, Zn, Ni, Co, Cr, Sb, Li, Be) have only a few ($< 10\%$) or no values below the chemical detection limit. All ten elements were log-transformed and standardized to reduce the influence of magnitudes. Their distributions had a fairly symmetrical shape after transformation, except for Au and Sb. Cr had a bimodal shape.

Correlation

The scatter diagrams of Au against each of the nine auxiliary elements are absolutely unstructured. As these diagrams represent only the 33% of the samples for which Au is above the detection limit, we calculated conditional means of Au with each of the auxiliary elements. The most interesting diagrams were found for Sb, As, Li and Pb. They correspond well to the values of the correlation coefficients with Au shown in Table 2. Although the correlation coefficients with Au are very weak, they confirm often-observed relationships (except for Li, which is not mentioned in the literature).

Bonnemaison (1986, p. 55) states: "it is possible, according to the relationships between the gold and the sulfide paragenesis, to establish semi-strategic geochemical tracers, such as As, which characterize the entire gold-bearing shear zone, and tactical tracers, such as Pb, which are characteristic specifically of the zones rich in gold." In that paper (Bonnemaison, 1986, p. 63), Sb is mentioned as a tactical or semi-strategic tracer, depending on the paragenesis.

Principal Components

Principal component analysis (PCA) of the correlation matrix (as described in Morrison, 1978) gave a first component (37% of the trace) contrasting {Ni, Co, Cr, Zn, Sb} against {Li, Be}. The second principal component (19%) associates {Li, Be} with Pb and to a lesser extent with As and Zn. Au is near the origin on the plane spanned by these two principal components.

The third component, F3 (15% of the trace), can be called the "gold factor," because of its compatibility with the metallogenetic model:

$$\text{F3} = \overbrace{(0.5\ \text{Au} + 0.5\ \text{As} + 0.4\ \text{Sb} + 0.2\ \text{Pb})}^{\text{positive contribution}}$$
$$- \underbrace{(0.3\ \text{Li} + 0.2\ \text{Be} + 0.1\ \text{Ni} + 0.2\ \text{Co} + 0.1\ \text{Cr} + 0.1\ \text{Zn})}_{\text{negative contribution}}$$

A proportional-circle map of the PCA gold factor is shown in Figure 1.

Cross variograms have been computed between the principal components and it turns out that these are not spatially orthogonal! As an example, the direct variograms of the third and fourth components are presented on Figure 2, as well as the cross variogram between these components. At short distances the correlation between the components is not zero, which means that they are not uncorrelated at a small scale. They become uncorrelated only at large distances when the samples are uncorrelated. This stems from the fact that apparently the correlation between the variables is different at small and large scales. Therefore, we shall perform a coregionalization analysis to separate small-scale and large-scale correlation and then cokrige the gold factor from a PCA of the small-scale variation, as explained in Wackernagel (1988).

Coregionalization Analysis

The experimental variograms have been fitted with a nested multivariate variogram model

$$\mathbf{G}(\mathbf{h}) = \underbrace{\mathbf{B}_0\ nug(\mathbf{h})}_{\text{pointwise}} + \underbrace{\mathbf{B}_1\ sph(\mathbf{h},\ 900m)}_{\text{local}} + \underbrace{\mathbf{B}_2\ sph(\mathbf{h},\ 3.5km)}_{\text{regional}}$$

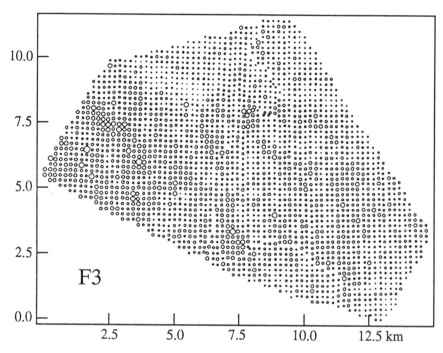

Figure 1: Map of the PCA gold factor with samples shown as circles proportional to the values.

where $\mathbf{G(h)}$ is the matrix of direct and cross variograms, $nug(\mathbf{h})$ is an elementary nugget-effect variogram function, and the $sph(\mathbf{h},\ a)$ are elementary spherical variogram models of range a. The coregionalization matrices $\mathbf{B}_0$, $\mathbf{B}_1$, $\mathbf{B}_2$ describe pointwise, local, and regional variances and covariances. Such a coregionalization model can be efficiently fitted using a least-squares algorithm proposed by Goulard (1989). The coregionalization parameters describing covariance b_{ij}^u and correlation r_{ij}^u with gold are shown in Table 3.

Cokriging the Short-Range Gold Factor

The first four eigenvalues λ_p^1 and the corresponding eigenvectors $\mathbf{q}_p^1$ of the coregionalization matrix $\mathbf{B}_1$ are listed in Table 4. The first eigenvector (32%) of $\mathbf{B}_1$ is obviously the "gold factor," roughly composed of this linear combination of elements:

Table 3: Coregionalization Parameters with Gold

		b_0	b_1	b_2	r	r_0	r_1	r_2
Au	As	.054	.128	.018	.16	.11	.67	.16
Au	Sb	.037	.037	.056	.27	.10	.28	.47
Au	Pb	.052	−.001	.051	.09	.08	.00	.41
Au	Zn	−.032	.002	−.032	.02	−.06	.01	−.26
Au	Ni	−.023	−.001	.002	.09	−.06	−.01	.01
Au	Co	−.050	.005	−.026	.02	−.13	.04	−.17
Au	Cr	−.034	.011	−.016	.06	−.08	.07	−.10
Au	Li	−.029	−.041	.026	−.12	−.06	−.32	.17
Au	Be	.005	.000	.028	−.07	.01	.00	.16

Table 4: First Four Eigenvalues of $\mathbf{B}_1$ with
Corresponding Eigenvectors

$\mathbf{B}_1$	λ_1^1	λ_2^1	λ_3^1	λ_4^1
	.47051	.37547	.26417	.13817
in %	31.7	25.3	17.8	9.3
	q_1^1	q_2^1	q_3^1	q_4^1
Au	.30169	.19525	−.32897	.19469
As	.68978	.30057	−.20394	−.05843
Sb	.37690	.26290	.03718	−.36576
Pb	.22630	.15707	.65253	.62538
Zn	−.04350	.39090	.41248	−.23186
Ni	−.27408	.44042	−.09429	.12446
Co	−.26167	.36442	−.21871	−.12078
Cr	−.28365	.50698	−.20899	.31373
Li	−.11825	.20287	.39046	−.50153
Be	.00000	.00000	.00000	.00000

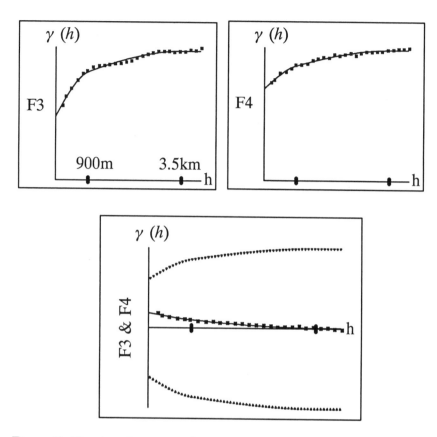

Figure 2: Direct variograms and cross variogram of the third and fourth principal components.

$$F1 \; = \; \overbrace{(0.7 \; As + 0.4 \; Sb + 0.3 \; Au + 0.2 \; Pb)}^{\text{positive contribution}}$$
$$- \underbrace{(0.3 \; Cr + 0.3 \; Ni + 0.3 \; Co + 0.1 \; Li + 0.04 \; Zn \,)}_{\text{negative contribution}} + 0.0 \; Be$$

The map of the gold factor from the factorial cokriging analysis (FCA) was cokriged with a neighborhood of 36 samples and is shown in Figure 3. Values below zero have been blanked out, as we are interested only in the high, positive values of the gold factor. The anomalies partly reflect sites where the Romans mined for gold. The L-shaped anomaly in the west highlights a geological structure which contains much As and Sb.

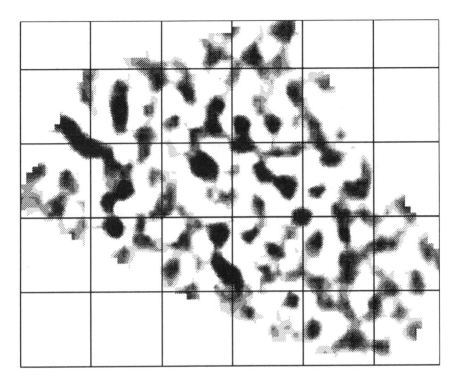

Figure 3: Map of the FCA gold factor, blanking out negative values.

Cokriging of Gold with its Tracers

For comparison we have cokriged the short-range structure of Au using the tracers As and Sb, because of their high positive correlation, and Li, for its high negative correlation (an anti-tracer, so to say). The resulting map of the short-range spatial component of gold is shown in Figure 4. The main anomaly, a little above the upper center of the map, is the former mine of Champvert which produced 2156 tonnes of quartz with 35 g/t Au during the period 1913–1949.

The map of the short-range gold factor is clearly more interesting than the map of the cokriging of the short-range gold component. This can be explained by the fact that the weights on the tracers are stronger in the factorial kriging than in the component kriging. The tracers contain interesting information about gold mineralization, while the data about gold itself, with many values below detectability, are extremely poor.

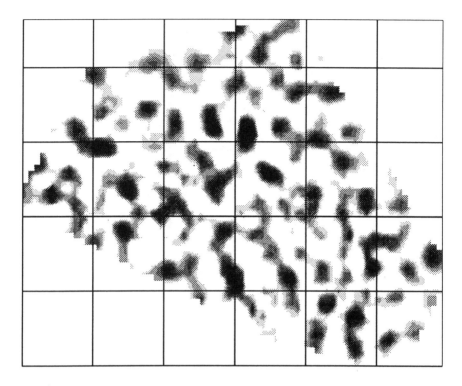

Figure 4: Map of cokriged short-range gold component, blanking out the negative values.

Conclusions

The main shortcoming of classical PCA applied to geochemical data is that the components (which are orthogonal in multivariate space) are not necessarily spatially orthogonal. This can be learned from the cross variogram between principal components shown in Figure 2. In such a situation, a coregionalization analysis based on a nested variogram model can help separate the different spatial scales of multivariate variation. We were thus able to define a "gold factor," based on variance and covariance at a local scale, which is consistent with the metallogenetic model of the shear zones in the Limousin. Cokriging of the gold factor resulted in a map which highlights interesting local anomalies. Cokriging of gold using the three tracers As, Sb, and Li as auxiliary variables gave less satisfactory results.

Looking at recent developments in computational statistics, it would be interesting to compare geostatistical analysis with the dynamic graphics techniques of Haslett *et al.* (1991). As a particular example, we would like to mention a striking metallogenetic coincidence in the case studies of Haslett *et al.* (1991) and Wackernagel and Butenuth (1989). The spatial and multivariate behavior of Cu, Zn, Pb shown by the two approaches is analogous in both papers, although the data came from very different regions.

Acknowledgments

The authors wish to thank the mining company COGEMA for permission to publish this case study. Joël Delair, Michel Normand and Guy Bourdin initiated the project.

References

Bonnemaison, M., 1986, Les "filons de quartz aurifère": Un cas particulier de shear zone aurifère: *Chronique de la Recherche Minière*, v. 482, p. 55–66.

Bonnemaison, M. and Marcoux, E., 1987, Les zones de cisaillement aurifères du socle hercynien français (avec discussion): *Chronique de la Recherche Minière*, v. 488, p. 29–42.

Bonnemaison, M. and Marcoux, E., 1990, Auriferous mineralization in some shear-zones: A three-stage model of metallogenesis: *Mineralium Deposita*, v. 25, p. 96–104.

Chenevoy, M., 1983, *Notice explicative de la feuille No. 713 'Châteauneuf-la-Forêt' à 1/50 000:* Service Géologique National, BRGM, Orléans, 48 pp.

Goulard, M., 1989, Inference in a coregionalization model, *in* Armstrong, M., (ed.), *Geostatistics:* Kluwer Academic Publisher, Amsterdam, p. 397–408.

Haslett, J., Bradley, R., Craig, P.S., Wills, G. and Unwin, A.R, 1991, Dynamic graphics for exploring spatial data, with application to locating global and local anomalies: *American Statistician*, v. 45, p. 234–242.

Morrison, D.F., 1978, *Multivariate Statistical Analysis:* McGraw–Hill, Singapore, 415 pp.

Wackernagel, H., 1988, Geostatistical techniques for interpreting multivariate spatial information, *in* Chung, C.F., *et al.*, (eds.), *Quantitative Analysis of Mineral and Energy Resources:* Reidel, Dordrecht, p. 393–409.

Wackernagel, H. and Butenuth, C., 1989, Caractérisation d'anomalies géochimiques par la géostatistique multivariable: *Jour. Geochemical Exploration*, v. 32, p. 437–444.

4

RECENT EXPERIENCES WITH PROSPECTOR II

Richard B. McCammon

Three recent case studies in which Prospector II was used illustrate a variety of constructive responses that contribute to regional mineral resource assessments. The case studies included a group of precious-metal vein deposits in the Quartzville Mining District in Oregon, United States; a stratabound gold-silver deposit in Manitoba, Canada; and an Archean tin deposit from Western Australia. In each case, the objective was to see how Prospector II would classify the deposit in terms of deposit models in the Cox-Singer compendium. The precious-metal vein deposits in the Quartzville Mining District were interpreted by Prospector II to be part of a larger system likely to contain porphyry copper deposits. The stratabound gold-silver deposit in Manitoba fit the description of the Homestake gold deposit model. The Archean tin deposit from Western Australia bore little resemblance to any of the tin deposit models in the Cox-Singer compendium.

Quantitative Assessments

In recent years, quantitative mineral resource assessments have gained recognition among land managers and national policymakers, who have found that numerical measures of potential mineral values are essential when considering alternative strategies. Such quantitative assessments allow land managers to plan optimum use of public lands and allow national policymakers to assess the need for securing long-term mineral supplies from international sources. In addition, quantitative assessments encourage the discovery and development of new deposits.

Significant advances have been made in developing new techniques for the quantitative assessment of metallic mineral resources (Drew *et al.*, 1986; Reed *et al.*, 1989). In large part, these techniques are based on an earlier method of regional mineral resource assessment proposed by Singer (1975) and subsequently applied to areas in Alaska. The technique is based on the size distribution of mineral deposits of specified geologic types and on the probability of deposit occurrence. This approach to the quantitative assessment of undiscovered mineral resources is being applied to many of the mineral resource assessments being carried out by the U.S. Geological Survey (USGS) (Singer and Cox, 1988).

Critical in this approach to quantitative assessment is the geologist's ability to relate the geologic environment in an area to specific deposit types. Mineral deposit models have been developed for this purpose that relate the common features of a specific deposit type to a particular geologic environment and are being used as the basis for quantitative mineral resource assessments. Efforts directed at model compilation began about 10 years ago within the USGS and culminated in *USGS Bulletin 1693* (Cox and Singer, 1986) which contains 89 descriptive models. The models were based on data from more than 3,900 individual deposits located in 110 different countries. Each model defines a set of attributes common to a particular type of mineral deposit. If the geologic environment of an area is similar to the geologic environment associated with a particular deposit model, then the area is considered to have the potential for deposits of the type represented by the model, even if no deposits of that type have been identified in the area. The tonnage and grade data compiled for the well-known, thoroughly explored or mined deposits used to define a model lead to the quantitative estimates. Relating the geologic environment to the appropriate set of deposit models helps ensure the overall accuracy and reliability of the final estimates.

Prospector II

In performing an assessment, the geologist faces the problem of deciding which of the 89 different deposit models, if any, best fit the data. The difficulty in making such a decision stems partly from the quality of the data collected for regional mineral resource assessments and partly from the level of experience of the geologist. The deposit models from which the geologist can choose represent a broad spectrum of mineralized systems. In fact, it would be an unusual geologist who had personal knowledge about all of the types of mineral deposits. For this reason, Prospector II was developed; it is to aid the geologist in deciding which of the 89 deposit models best fit the data. As the successor to the original Prospector system (Duda, 1980), Prospector II helps the geologist judge how nearly a model matches a given set of data by calculating a score that expresses the degree of similarity between the model's parameters and the data.

Although models in Prospector II are patterned after those in *USGS Bulletin 1693*, they are represented as a set of weighted attributes rather than a narrative description. A pair of numerical weights (values) is associated with

Table 1: Verbal Descriptions for States of Presence–Absence

State	Numerical value	Verbal description
	Degree of sufficiency	
	5	Very highly suggestive
	4	Highly suggestive
Presence	3	Moderately suggestive
	2	Mildly suggestive
	1	Weakly suggestive
	Degree of necessity	
	−1	Infrequently present
	−2	Occasionally present
Absence	−3	Commonly present
	−4	Most always present
	−5	Virtually always present

each model attribute and represents different measures of the attribute's relative importance, depending on whether the attribute is present or absent in the data. The first measures the *degree of sufficiency* of the attribute and reflects how much importance is placed on its presence. The second measures the *degree of necessity* of the attribute and reflects how much importance is placed on its absence. Measures of degrees of sufficiency and necessity range from +5 to −5. The verbal descriptions associated with these values are given in Table 1. A portion of the descriptive model for tin greisen deposits from Reed (1986) is shown in Table 2 and for comparison, the equivalent numerical model for tin greisen deposits is shown in Table 3.

In an ideal situation, the attributes of the individual deposits used to define a model should match all of the attributes of that model. This is rarely the case, however, and a model is defined by the common attributes of the individual deposits rather than by all attributes. Moreover, different deposit models are characterized by different numbers of common attributes. So that the measures of the degrees of sufficiency and necessity are comparable across different models, these measures must be transformed into scores that correspond to the major headings used to define the different models. The transformations from measures to scores are shown in Table 4. The scores are assigned so that the individual deposits used to define a model achieve their highest score when they are compared to the model of the type that they represent (McCammon, 1992). Thus, differences in the number of attributes within major headings and the differences in the relative importance of the major headings in defining a model were taken into account. For example,

Table 2: A Portion of the Descriptive Model for Tin Greisen
Deposits Taken from Reed (1986)

Description:	Disseminated cassiterite and cassiterite-bearing veinlets, stockworks, lenses, pipes, and breccia in greisenized granite.
Geologic ages:	May be any age; tin mineralization temporally related to later stages of granitoid emplacement.
Rock types:	Specialized biotite ± muscovite leucogranite (S-type); distinctive accessory minerals include topaz, fluorite, tourmaline, and beryl.
Form–Structure:	Exceedingly varied, the most common being disseminated cassiterite in massive greisen, and quartz veinlets and stockworks; less common are pipes, lenses, and tectonic breccia.
Alteration:	Incipient greisen (granite); muscovite ± chlorite, tourmaline, and fluorite. Greisenized granite; quartz-muscovite-topaz-fluorite ± tourmaline. Massive greisen; quartz-muscovite-topaz ± fluorite ± tourmaline.
Minerals:	General zonal development of cassiterite + molybdenite, cassiterite + molybdenite + arsenopyrite + beryl, wolframite + beryl + arsenopyrite + bismuthinite, Cu-Pb-Zn sulfide minerals + sulphostannates, quartz veins ± fluorite, calcite, pyrite.
Geochemical–Elements:	Specialized granites enriched in Sn, F, Rb, Li, Be, W, Mo, Pb, B, Nb, Cs, U, Th, Hf, Ta, and most REE's.

Table 3: A Portion of the Numerical Model for Tin Greisen
Deposits in Prospector II

Description:	Disseminated cassiterite and cassiterite-bearing veinlets, stockworks, lenses, pipes, and breccia in greisenized granite.
Geologic ages:	Precambrian OR Phanerozoic $(5 -5)$.[*]
Rock types:	Felsic plutonic $(5 -5)$, granite $(5 -5)$, leucogranite $(4 -4)$, muscovite leucogranite $(3 -2)$, biotite-leucogranite $(3 -2)$.
Form–Structure:	Greisen, veinlets, stockwork.
Alteration:	Greisenization $(5 -2)$, albitization $(5 -2)$, tourmalinization $(3 -2)$.
Minerals:	Cassiterite $(4 -5)$, molybdenite $(4 -5)$, arsenopyrite $(3 -5)$, topaz $(4 -2)$, tourmaline $(4 -2)$, beryl $(2 -4)$, wolframite $(2 -3)$, bismuthinite $(2 -2)$, fluorite $(4 -3)$, calcite $(1 -3)$, pyrite $(2 -4)$.
Geochemical–Elements:	Sn $(4 -5)$, F $(5 -5)$, B $(5 -4)$, Mo $(2 -5)$, Rb $(2 -4)$, Cs $(2 -4)$, Be $(2 -3)$, REE $(2 -4)$, U $(2 -4)$, Th $(2 -4)$, Nb $(2 -4)$, Ta $(2 -4)$, Li $(2 -4)$, W $(2 -3)$, As $(2 -4)$, Bi $(2 -3)$.

[*]The pair of values in parentheses after each attribute
refers to the measure of the degree of sufficiency and the
degree of necessity, respectively. The attributes listed under
the headings Description and Form–Structure are not used in
calculating scores. Unlisted attributes have no significance
with respect to the model.

Table 4: Numerical Values and Associated Scores for States of Presence–Absence

Value	Presence						Absence				
	5	4	3	2	1	0	-1	-2	-3	-4	-5
Age:	100	40	40	40	40	0	-100	-100	-100	-100	-100
Rk:	75	60	45	30	15	0	-5	-10	-45	-60	-75
Alt:	400	300	200	100	50	0	-2	-10	-100	-200	-400
Min:	75	60	45	30	15	0	0	-5	-10	-30	-75
Gx:	75	60	45	30	15	0	0	-5	-10	-30	-75
Gp:	250	150	50	25	10	0	-10	-50	-100	-200	-250
Dep:	400	320	200	150	75	0	-50	-100	-200	-300	-400

Age: geologic ages   Gx: geochemical elements
Rk: rock types   Gp: geophysical signatures
Alt: alteration   Dep: associated deposits
Min: minerals

only a few rock types usually are needed to characterize a model with respect to rock type whereas several minerals may be needed to characterize a model in terms of mineralogy. Though rock type and mineralogy are important attributes, the presence of one type of alteration may be all that is necessary to narrow the selection to a single model. Attributes listed under the major heading "Form–Structure" are not used for scoring, but rather as the basis for selecting among a set of deposit models in a given situation. These attributes describe the morphology of deposits, and ordinarily morphology will be well recognized only after exploration is far advanced. For further details about the Prospector II system, the reader is referred to McCammon (1984, 1989, 1990, 1991, 1992).

In this paper, three case studies are presented in which Prospector II was used to determine which deposit model best fit a particular deposit description derived from *USGS Bulletin 1693*. The three case studies illustrate how Prospector II served as a consultant and how the advice offered by Prospector II bears on deposit model classification in general and regional mineral resource assessments in particular.

Case Studies

The first case study resulted from a request in 1989 from Mr. Steven R. Munts, a consultant, who was interested in how Prospector II would classify deposits in the Quartzville Mining District located in the Western Cascades of west-central Oregon. The district was one of the early gold producers in Oregon (Callaghan and Buddington, 1938). Deposits contain gold and silver occurring in veins and silicified shear zones in rhyolite, andesite, and tuff, associated with dacite porphyry dikes and plugs. The historic production of

Table 5: Description of Gold–Silver Veins in the Quartzville
Mining District, Western Cascades, Oregon,
Submitted to Prospector II

Geologic ages:	Tertiary.
Rock types:	Rhyodacite, basalt, rhyolite, andesite, dacite, diorite, quartz diorite, granodiorite, aplite.
Form– Structure:	Veins, veinlets.
Alteration:	Silicification, propylitic, argillic, phyllic, sericite.
Minerals:	Pyrite, chalcopyrite, gold, galena, sphalerite, stibnite, barite, bornite, magnetite.
Geochemical– Elements:	Zn, Cu, Pb, Ag, Au, As.
Geophysical signatures:	Magnetic low.

gold in the district is about 7 million ounces. The deposits are thought to
be part of an altered caldera, and the question is whether there is evidence
of other types of deposits. The description supplied by Munts is shown in
Table 5 and consists of lists of attributes for major headings that are used
in defining deposit models. With this description, Prospector II produced
the results shown in Table 6. The first choice of Prospector II, based on the
description provided, was the porphyry copper deposit model.

Independently and at a later date, a team of USGS geologists evaluated the undiscovered mineral resources in northwestern California, western
Oregon, and western Washington and concluded that the area including
the Quartzville Mining District was permissive for the occurrence of undiscovered porphyry copper deposits (Diggles *et al.*, 1991). Their conclusions
were based on the available geologic, geochemical, geophysical, and mineral
resource data. Geophysical data were interpreted as indicating the presence of buried plutons in and around the Quartzville Mining District. In
this instance, Prospector II's evaluation was verified by a team of USGS
geologists.

The second case study involved a letter from Dr. Mark Fedikow, Geological Services Branch, Manitoba Department of Energy and Mines, Canada,
who was interested in how Prospector II would classify the MacLellan (Agassiz) gold-silver deposit near Lynn Lake in northern Manitoba. This deposit
is the largest gold deposit in a zone 70 km long by 1 km wide in the northern
belt of the Aphebian Lynn Lake greenstone belt (Fedikow *et al.*, 1989). The
MacLellan deposit is characterized by disseminated to solid gold-bearing sulfide minerals that occur within a picritic basalt, siliceous layers, and quartz
veins. The deposit has been strongly deformed with accompanying sulfide
deposition and gold mobilization. The deposit also is enriched in antimony,

Table 6: Results Obtained from Prospector II
for Description in Table 5

The calculated scores for the top six deposit models are:	

1	2108	Porphyry copper
2	2103	Porphyry molybdenum low-fluorine
3	1258	Porphyry copper-molybdenum
4	1165	Polymetallic veins
5	1055	Low-sulfide gold quartz veins

The attributes in boxes are explained by the porphyry copper model.

Geologic ages:	Tertiary .
Rock types:	Rhyodacite, basalt, rhyolite, andesite, dacite, diorite, quartz diorite, granodiorite , aplite.
Form– Structure:	Veins , veinlets .
Alteration:	Silicification, propylitic , argillic , phyllic , sericite .
Minerals:	Pyrite , chalcopyrite , gold, galena, sphalerite, stibnite, barite, bornite , magnetite .
Geochemical– Elements:	Zn, Cu , Pb, Ag , Au , As.
Geophysical signatures:	Magnetic low.

arsenic, and mercury. The description supplied by Fedikow is shown in Table 7 and Prospector II produced the results shown in Table 8.

It happens that this deposit was one of 116 deposits used to construct the Homestake gold deposit model (Berger, 1986; Mosier, 1986). Thus, Prospector II demonstrated that it can classify correctly a deposit that was used in constructing the original Homestake gold deposit model.

The third case study resulted from a letter received from Dr. Eric Grunsky, Division of Exploration Geoscience, CSIRO, Australia, who was interested in how Prospector II would classify the Greenbushes tin deposit in Western Australia (Smith *et al.*, 1987). The deposit is part of the Greenbushes tin-tantalum pegmatite district, a mineralized pisolitic laterite duricrust enriched in arsenic, tin, beryllium, and antimony. Historic production amounts to 30 to 40 million tonnes of pegmatite having an average grade of 220 ppm tin, 25 ppm niobium, and 30 ppm tantalum. The question was whether this deposit fit any of the deposit models in *USGS Bulletin 1693*. Grunsky's description is shown in Table 9; Prospector II results are shown in Table 10. There is little similarity between the models and the data. The first choice, the tin-greisen deposit model, has only rock types and form–structure in common with the data. In this instance, it is safe to conclude that a new model for tin-bearing pegmatite deposits should be considered.

Table 7: Description of MacLellan (Agassiz) Gold–Silver Deposit
near Lynn Lake, Manitoba, Canada,
Submitted to Prospector II

Geologic ages:	Archean.
Rock types:	Komatiite, iron formation, graywacke, mafic metavolcanic rocks.
Form–Structure:	Stratiform.
Alteration:	Silicification, propylitic, argillic, phyllic, sericite.
Minerals:	Pyrite, pyrrhotite, arsenopyrite, sphalerite, galena, chalcopyrite, magnetite, tetrahedrite.
Geochemical–Elements:	Au, As, Sb, Cu, Pb, Zn, Fe, Mn, K, Rb, Co, Ni, Ti,* Na.*
Geophysical signatures:	Induced polarization anomaly, electromagnetic anomaly.

*Anomalously low in concentration

Table 8: Results Obtained from Prospector II
for Description in Table 7

The calculated scores for the top five deposit models are:

1   575 Homestake gold
2   470 Blackbird cobalt-copper
3   440 Kuroko massive sulfide
4   335 Cyprus massive sulfide
5   320 Besshi massive sulfide

The attributes in boxes are explained by the Homestake gold model.

Geologic ages:	[Archean] .
Rock types:	[Komatiite] , [iron formation] , graywacke, [mafic metavolcanic rocks] .
Form–Structure:	[Stratiform] .
Alteration:	Silicification, propylitic, argillic, phyllic, sericite.
Minerals:	[Pyrite] , [pyrrhotite] , [arsenopyrite] , [sphalerite] , galena, chalcopyrite, [magnetite] , tetrahedrite.
Geochemical–Elements:	[Au] , [As] , [Sb] , Cu, Pb, Zn, [Fe] , Mn, K, Rb, Co, Ni, Ti,* Na.*
Geophysical signatures:	Induced polarization anomaly, electromagnetic anomaly.

*Anomalously low in concentration

Table 9: Description of Greenbushes Tin Deposit in
Western Australia Submitted to Prospector II

Geologic ages:	Archean.
Rock types:	Amphibolite, granite gneiss.
Form–Structure:	Pegmatite, laterite.
Minerals:	Cassiterite, beryl, tourmaline, spodumene.
Geochemical–Elements:	Sn, As, Sb, Ta, Be, Bi, Li, Nb.

Table 10: Results Obtained from Prospector II
for Description in Table 9

The calculated scores for the top five deposit models are:

1 465 Tin greisen
2 390 Tin veins
3 355 Tungsten veins
4 300 Tin polymetallic veins
5 300 Alluvial-placer tin

The attributes in boxes are explained by the tin greisen model.

Geologic ages:	Archean .
Rock types:	Amphibolite, granite gneiss.
Form–Structure:	Pegmatite, laterite.
Minerals:	Cassiterite , beryl , tourmaline , spodumene.
Geochemical–Elements:	Sn , As , Sb, Ta , Be , Bi , Li, Nb .

Conclusions

In each of these three case studies, Prospector II was used as a computer-based consultant to determine, among other things, how well the description of the deposit matched the deposit models stored in the knowledge base, which parts of the data were explained by a model, and whether a new deposit model was needed. Prospector II has proven to be a valuable tool for evaluating the available data, allowing for the rapid comparison of a large number of deposit models, and focusing on regional mineral resource assessment activities that involve identifying the appropriate set of deposit models. In future years, I anticipate that Prospector II will expand its coverage of different deposit types, address more vigorously the relation between deposits and lithotectonic settings, and evolve into a self-perpetuating system to use as a teaching tool for less experienced geologists and to assist in the evaluation of mineral-resource potential in given areas.

References

Berger, B.R., 1986, Descriptive model of Homestake–Au, in Cox, D.P. and Singer, D.A., (eds.), Mineral deposit models: *USGS Bulletin 1693*, U.S. Geological Survey, p. 244.

Callaghan, E. and Buddington, A.F., 1938, Metalliferous mineral deposits of the Cascade Range in Oregon: *USGS Bulletin 893*, U.S. Geological Survey, 141 pp.

Cox, D.P. and Singer, D.A., (eds.), 1986, Mineral deposit models: *USGS Bulletin 1693*, U.S. Geological Survey, 379 pp.

Diggles, M.E., (ed.), 1991, Assessment of undiscovered porphyry copper deposits within the range of the northern spotted owl, northwestern California, western Oregon, and western Washington: *Open-File Report 91–377*, U.S. Geological Survey, 58 pp.

Drew, L.J., Bliss, J.D., Bowen, R.W., Bridges, N.J., Cox, D.P., DeYoung, J.H., Jr., Houghton, J.C., Ludington, S., Menzie, W.D., Page, N.J., Root, D.H. and Singer, D.A., 1986, Quantitative estimation of undiscovered mineral resources—A case study of U.S. Forest Service wilderness tracts in the Pacific Mountain system: *Economic Geology*, v. 81, p. 80–88.

Duda, R.O., 1980, *The Prospector System for Mineral Exploration:* Final Report SRI Project 8172, Menlo Park, California, 120 pp.

Fedikow, M.A.F., Parbery, D. and Ferreira, K.J., 1989, Agassiz metallotect—A regional metallogenetic concept, Lynn Lake Area: Mineral Deposit Thematic Map Series MAP 89-1: Geological Services Branch, Department of Energy and Mines, Winnipeg, Manitoba, 1 map sheet.

McCammon, R.B., 1984, Recent developments in PROSPECTOR and future expert systems in regional resource evaluation: *IEEE Proceedings Pecora IX Spatial Information Technologies for Remote Sensing Today and Tomorrow*, Sioux Falls, South Dakota, p. 243–248.

McCammon, R.B., 1989, Prospector II—The redesign of Prospector: *AI Systems in Government*, March 27–31, 1989, Washington, D.C., p. 88–92.

McCammon, R.B., 1990, Maintaining Prospector II as a full-sized knowledge-based system: *AISIG '90 Research Workshop on Full-Sized Knowledge Based Systems in Government*, May 1990, Washington, D.C., 10 pp.

McCammon, R.B., 1991, Prospector II: An expert system for mineral resource assessment on public lands: *Proceedings, Resource Technology 90, 2nd International Symposium on Advanced Technology in Natural Resource Management*, November 12–15, 1990, Washington, D.C., p. 718–726.

McCammon, R.B., 1992, Numerical mineral deposit models, in Bliss, J.D., (ed.), New developments in mineral exploration and resource assessment: *USGS Bulletin 2004*, U.S. Geological Survey, p. 6–12, 64–167.

Mosier, D.L., 1986, Grade and tonnage model of Homestake–Au, in Cox, D.P. and Singer, D.A., (eds.), Mineral deposit models: *USGS Bulletin 1693*, U.S. Geological Survey, p. 245–247.

Reed, B.L., 1986, Descriptive model of Sn greisen deposits, in Cox, D.P. and Singer, D.A., (eds.), Mineral deposit models: *USGS Bulletin 1693*, U.S. Geological Survey, p. 70.

Reed, B.L., Menzie, W.D., McDermott, M., Root, D.H., Scott, W.A. and Drew, L.J., 1989, Undiscovered lode tin resources of the Seward Peninsula, Alaska: *Economic Geology*, v. 84, p. 1936–1947.

Singer, D.A., 1975, Mineral resource models and the Alaskan Mineral Resource Assessment Program, in Vogely, W.A., (ed.), *Mineral Materials Modeling—A State-of-the-Art Review:* Johns Hopkins Press, Baltimore, p. 370–382.

Singer, D.A. and Cox, D.P., 1988, Applications of mineral deposit models to resource assessments: *Yearbook Fiscal Year 1987*, U.S. Geological Survey, p. 55–57.

Smith, R.E., Perdrix, J.L. and Davis, J.M., 1987, Dispersion into pisolitic laterite from the Greenbushes mineralized Sn–Ta pegmatite system, Western Australia: *Jour. Geochemical Exploration*, v. 28, p. 251–265.

5

CORRESPONDENCE ANALYSIS IN HEAVY MINERAL INTERPRETATION

J. Tourenq, V. Rohrlich and H. Teil

Correspondence analysis, a non-parametric principal component analysis, has been used to analyze heavy mineral data so that variations between both samples and minerals can be studied simultaneously. Four data sets were selected to demonstrate the method. The first example, modern sediments from the River Nile, illustrates how correspondence analysis brings out extra details in heavy mineral associations. The other examples come from the Plio-Quaternary "Bourbonnais Formation" of the French Massif Central. The first data set demonstrates how the principal factor plane (with axes 1 and 2) highlights relationships between geographical position and the predominant heavy mineral association (metamorphic minerals and zircon), suggesting the paleogeographic source. In the second set, the factor plane of axes 1 and 3 indicates a subdivision of the metamorphic mineral assemblage, suggesting two sources of metamorphic minerals. Finally, outcrop samples were projected onto the factor plane and reveal ancient drainage systems important for the accumulation of the Bourbonnais sands.

Interpreting Heavy Minerals in Sediments

Statistical methods used in interpreting heavy minerals in sediments range from simple and classical methods, such as calculation of means and standard deviations, to the calculation of correspondences and variances. Use of multivariate methods is increasingly frequent (Maurer, 1983; Stattegger, 1986; 1987; Delaune *et al.*, 1989; Mezzadri and Saccani, 1989) since the first studies of Imbrie and van Andel (1964). Ordination techniques such as principal component analysis (Harman, 1961) synthesize large amounts of data and extract the most important relationships. We have chosen a non-parametric form of principal component analysis called correspondence analysis. This technique has been used in sedimentology by Chenet and Teil (1979) to investigate deep-sea samples, by Cojan and Teil (1982) and Mercier *et al.* (1987) to define paleoenvironments, and by Cojan and Beaudoin (1986) to show paleoecological control of deposition in French sedimentary basins. Correspondence analysis has been used successfully to interpret heavy mineral data (Tourenq *et al.*, 1978a, 1978b; Bolin *et al.*, 1982; Tourenq, 1986, 1989; Faulp *et al.*, 1988; Ambroise *et al.*, 1987). We provide examples of different situations where the method can be applied.

Methodology

We will not present the mathematical and statistical procedures involved in correspondence analysis, but refer readers to Benzécri *et al.* (1973), Teil (1975), Greenacre (1984), and Lebart *et al.* (1984). Briefly, correspondence analysis is a data-reducing technique which can be applied to tables of positive numbers. In our data, the numbers represent i samples of j abundances of heavy minerals. No data transformation takes place because correspondence analysis computes a bi-dimensional probability matrix from the initial data set (Jambu and Lebaux, 1983), in which each value is divided by the total sum. Thus, the individuals have proportional weights and the "profile" of each is considered.

Each sample can be visualized as a point in j-dimensional space, and in this space, the most significant components (axes) can be extracted by diagonalization of the probability matrix. Generally, the first few axes represent the maximum amount of variability and are thus studied in detail. The other axes, however, reflect minor proportions and usually are considered redundant, although in certain situations they can pinpoint a specific relationship. The samples can be projected onto three or more planes (axes 1 and 2; 1 and 3; 2 and 3) together with the variables, which is one advantage of correspondence analysis. For example, it is possible to say that a sample near the projection of zircon (*e.g.*, ZIR in Fig. 5a) has a relatively high zircon content, as do other nearby samples (on a correspondence analysis graph, proximity means similarity).

Interpretation by Correspondence Analysis

Many sediments such as fluviatile and turbidite deposits have the same mineral composition, but in varying proportions. Correspondence analysis reduces the number of variables, allowing easier understanding of underlying regularities. Mineral abundance usually has a non-normal (non-Gaussian) distribution which excludes the use of many multivariable methods, but

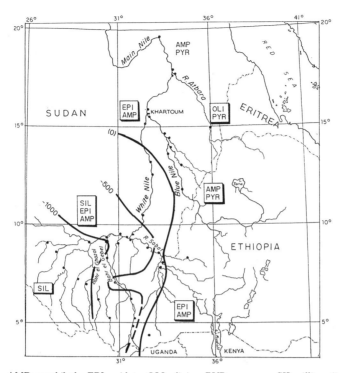

AMP: amphibole; EPI: epidote; OLI: olivine; PYR: pyroxene; SIL: sillimanite

Figure 1: Map of Nile basin showing sampling localities (from Shukri, 1950). Contour lines represent factor 1 scores.

correspondence analysis is a non-parametric technique. Use of correspondence analysis helps to minimize the closure effect in percentage data. In addition, "supplementary" samples can be projected onto a set of factor planes without taking part in the actual factor computations.

We have chosen four examples in order to present various facets of correspondence analysis and to show how this method can be applied to heavy mineral data interpretation. In the first example, the evolution of heavy mineral associations in Nile River sediments are analyzed with greater precision than would otherwise be possible. The other three examples are from a continental Plio-Quaternary formation in Central France and show the use of the principal plane (1 *vs.* 2 axes), the 1 *vs.* 3 axes plane, and "supplementary" samples.

Modern Sediments of the River Nile

Shukri (1950, 1951) investigated heavy minerals in modern sediments of the River Nile and its tributaries (Fig. 1), and his data were used by Morton (1985) in provenance studies (Fig. 2). Morton, on the basis of six significant

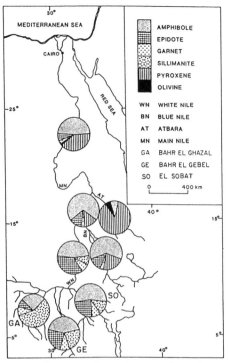

Figure 2: Heavy minerals in modern River Nile sediments. Note marked effects on composition caused by confluence of Blue Nile and White Nile and by confluence of Nile and Atbara (from Morton, 1985).

minerals, calculated the average heavy mineral composition for each tributary: the Bahr el Ghazal and its tributaries were dominated by sillimanite; the Bahr el Gebel by sillimanite, amphibole, and epidote; the Sobat by epidote and amphibole; the Blue Nile by amphibole; and the Atbara by pyroxene and olivine. The White Nile and the Main Nile above the confluence with the Atbara were characterized by amphibole and epidote, while the Main Nile below the confluence with the Atbara showed an increase in the proportion of pyroxene.

Correspondence analysis was applied to Shukri's data for about 50 specimens with 12 heavy minerals. The factor contributions and the projection on the principal plane show that pyroxene, amphibole, and olivine are opposed to sillimanite, garnet, and epidote on the first factor (Fig. 3). Factor 2 shows the opposition of amphibole and epidote to pyroxene and sillimanite. Other minerals have only a minimal effect.

On the principal plane, samples show a distinct cluster of points between the Bahr el Ghazal, where sillimanite predominates, and the Atbara, where pyroxene and olivine predominate. In the Bahr el Gebel and the Sobat,

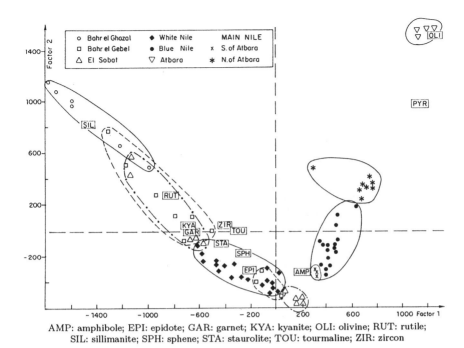

AMP: amphibole; EPI: epidote; GAR: garnet; KYA: kyanite; OLI: olivine; RUT: rutile;
SIL: sillimanite; SPH: sphene; STA: staurolite; TOU: tourmaline; ZIR: zircon

Figure 3: Correspondence analysis of heavy minerals, recent Nile sediments.

analysis of all specimens shows a distribution with two distinct clusters for each tributary. The first cluster is between the sillimanite and garnet poles (the clusters of Bahr el Gebel and Sobat occur together). The second cluster occurs between the epidote and amphibole poles (in this case the clusters of each tributary are also very close together). The Blue Nile has a distinct position on the positive side of factor 1. Specimens from the Main Nile above the confluence with the Atbara are included within this cluster, while samples below the confluence shift toward the pyroxene pole.

Minerals significant in specific clusters are indicated on the location map (Fig. 1), together with factor lines to indicate the area with factor scores between 0 and −1000, and the area with scores below −1000. As in Figure 3, samples of the Gebel and Sobat are variable. Samples from the western part, Ghazal, and lower reaches of Bahr el Gebel have strongly negative scores because of their high sillimanite content. However, those from the rest of the Gebel and the western part of the Sobat are less negative. Eastern el Sobat samples have positive factor 1 scores.

The variations highlighted by factor 1 result from sillimanite derived from graphite and sillimanite-bearing gneiss in the S–SW sector of Sudan (Yassin *et al.*, 1984) and the relative increase in amphiboles derived from the Ethiopian highlands and basaltic source rocks for sediments of the Atbara. Correspondence analysis helps give a more general meaning to mineralogical zones which extend beyond each tributary. In particular, the White Nile is much more dependent on sediment supply coming from the Sobat region to the SSE, rather than a supply coming from the Bahr el Gebel and the Bahr el Ghazal to the SSW. Whether the predominant influence is linked to the quantity of sediment transported, to the quantity of minerals, or to the fluvial mobility of certain minerals remains to be resolved.

Bourbonnais Formation

Geological context. The sands and clays of the Bourbonnais Formation are Reuverian to Pretiglian in age (Upper Pliocene) and crop out from south of Vichy to north of Nevers along the banks of the Loire and Allier rivers (Fig. 4), with a maximum thickness of 50 m. They consist of three units, the lower sands, black clays, and upper sands. Grain size in the two sands decreases from the base upwards. Deposition took place in a subsiding basin during a short distensive phase. Sediments are essentially fluvial in origin, but may also be fluvio-lacustrine; some volcanic minerals (such as volcanic zircons) may have an eolian origin.

Data set. Eleven heavy minerals from hundreds of samples (mostly from wells) were identified and counted (Tourenq, 1989): tourmaline, zircon, staurolite, kyanite, andalusite, sillimanite, garnet, titaniferous minerals (rutile, anatase, and brookite: RAB), sphene, hornblende (and epidote), and monazite. Histograms were produced for each mineral. Most have a non-normal distribution.

Example 1: Use of the principal plane (1 *vs.* 2). Ninety samples from 12 wells (Section I in Figs. 4 and 5b) were analyzed for heavy minerals by Tourenq (1989). It is difficult to obtain an overall graphical representation for such a large number of samples using classical methods; therefore, average values for each well were used even though data are not always homogeneous throughout a well. In Table 1 mean compositions of each well are arranged from west to east, showing that zircon values increase eastward and staurolite decreases. Other relationships, however, are not easy to identify. Multivariate methods can be used to obtain a more precise interpretation.

All samples (not the well averages) were analyzed and projected onto the principal factor plane, but only four wells are retained for illustration (Fig. 5a). Axes 1 and 2 represent 59% of the variability. Axis 1 opposes zircon to metamorphic minerals: staurolite, kyanite, andalusite, sillimanite, and garnet. Zircon is the predominant influence on this axis (contributing 51.8%), and is highly correlated with it (0.95). The second axis opposes staurolite, zircon, tourmaline, and monazite to rutile, anatase, brookite (RAB), sphene, and hornblende.

On Figure 5, wells are represented by closed curves which envelop the cluster of points belonging to each well. The relationship between the geographical positions of these wells and their place on the principal plane can

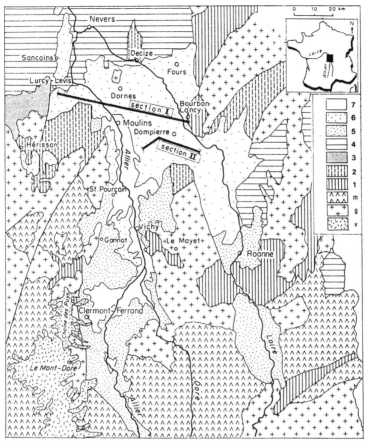

v: volcanic; g: granitic; m: metamorphic; 1: Devonian and Carboniferous; 2: Permian; 3: Triassic; 4: Jurassic; 5: Eocene–Oligocene; 6: Pliocene, Bourbonnais Fm.; 7: alluvium

Figure 4: Geological environment of sands and clays of Bourbonnais Formation showing Sections I and II (after Tourenq, 1989).

be seen clearly. The most easterly wells (SR, RA, and MONTAMB) are situated near the zircon side of the diagram, whereas the western well LM occurs on the other side. Other wells are intermediate both on the diagram and geographically. It is therefore possible to make a distinction between different sources of materials from the west metamorphic zone and from the east crystalline (zircon) zone. The middle represents a mixture from the two sources. The second axis can be interpreted as reflecting the volcanic sources of sphene and hornblende.

Example 2: Use of the 1 *vs.* 3 factor plane. Eighty-eight samples from eight wells along Section II (Fig. 5b) were analyzed for heavy minerals (Tourenq, 1989). Wells are situated south of the previous wells in the center of the

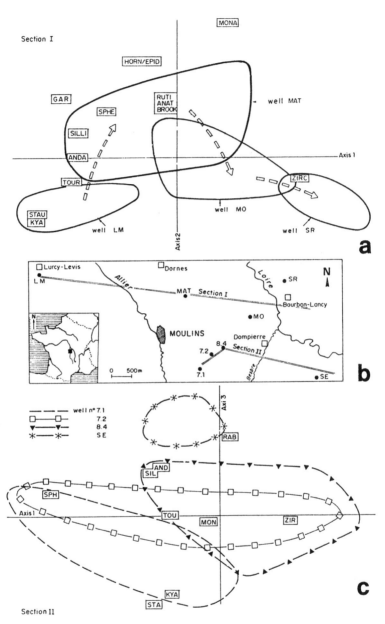

ANDA: andalusite; EPID: epidote; GAR: garnet; HORN: hornblende; KYA: kyanite; MONA: monazite; RAB: rutile, anatase, brookite; SILLI: sillimanite; SPHE: sphene; STAU: staurolite; TOUR: tourmaline; ZIR: zircon

Figure 5: Correspondence analysis along axes 1 and 2 for Section I (a) and along axes 1 and 3 for Section II (c); well location (b).

Table 1: Mean Composition of Heavy Minerals (Based on Calculation
of Number of Grains) in Samples from Wells from Section I
(Arranged from West to East)

	TOU	ZIR	RAB	STA	AND	GAR	KYA	SIL	SPH	HOR	MON
WELL											
West											
LM	27	18	8	19	14	0	4	8	1	0	1
FU	18	26	5	22	7	0	4	10	5	1	2
SL/LH	27	17	7	20	11	0	1	8	7	0	2
LMO	18	21	9	10	9	3	1	7	11	5	6
DBC	23	25	9	14	6	0	0	7	9	1	4
RB	18	19	6	19	10	2	1	6	17	2	1
MAT	14	30	19	5	7	1	1	8	7	3	5
RAT	9	47	12	2	7	0	0	5	3	0	15
SAB/ROY	12	39	17	3	9	0	0	6	7	1	7
MO	23	45	6	5	5	0	1	4	6	2	3
RA	2	90	5	0	1	0	0	0	1	0	1
MONTAMB	4	87	5	1	2	0	0	0	0	1	0
SR	10	74	11	0.5	1	0	0	0.5	1	2	0
East											

AND: andalusite; GAR: garnet; HOR: hornblende; KYA: kyanite; MON: monazite; RAB: rutile,
anatase, brookite; SIL: sillimanite; SPH: titanite; STA: staurolite; TOU: tourmaline; ZIR: zircon

basin. No conclusions can be drawn from the mean mineral composition of
each well (Table 2). Therefore, correspondence analysis was used to investigate any hidden relationships. Samples projected onto the principal plane
(1 *vs.* 2) show no pattern, so the 1 *vs.* 3 factor plane was investigated.

As in the previous example, the first factor contrasts zircon to sphene and
other minerals. The third factor explains only 13% of the variance, but sub-
divides the metamorphic minerals into two pairs: andalusite–sillimanite and
staurolite–kyanite. Thus, projection onto the 1 *vs.* 3 factor plane produces a
pattern (Fig. 5c) in which the cluster position can be related to geographical
location of the wells. The more easterly wells (illustrated only by the SE
well on the correspondence analysis graph) are situated along the third axis,
near the sillimanite–andalusite and RAB variables. This suggests a possible
source of metamorphic minerals from the east, dominated by sillimanite and
andalusite. On the first axis, sphene (from volcanism at Mont-Dore situated
to the southwest of the Bourbonnais Formation, Fig. 4) has a local effect
and influences the western wells (7.1 and 7.2).

Example 3: Use of "supplementary" samples. Additional samples from
outcrops (Fig. 6a) were analyzed (Tourenq, 1989) in order to increase the
study area. These are considered as "supplementary" samples and are projected onto the axial planes calculated in Example 2. Projection onto the
1 *vs.* 3 factor plane (Fig. 6a) shows that these form two distinct clusters:
one characterized by sphene and staurolite–kyanite, the other by zircon,

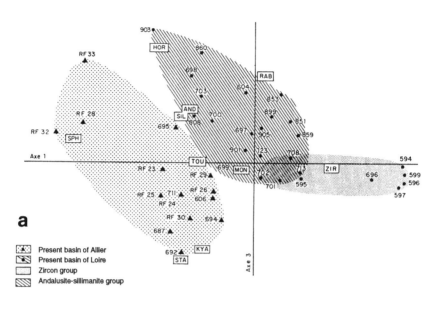

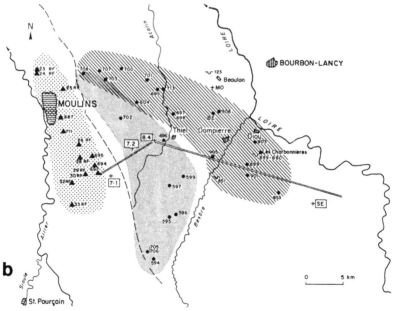

Figure 6: (a) Correspondence analysis along axes 1 and 3 for outcrop samples in wells in Section II. (b) Geographic distribution of zircon and andalusite–sillimanite groups in present basin of River Loire and staurolite and sphene group in present basin of River Allier.

Table 2: Mean Composition of Heavy Minerals (Based on Calculation
of Number of Grains) in Samples from Wells from Section II
(Arranged from West to East)

	TOU	ZIR	RAB	STA	AND	GAR	KYA	SIL	SPH	HOR	MON
WELL											
West											
BEF/EBE	9	50	6	16	4	4	1	3	7	0	3
7.1	19	24	4	24	10	1	3	5	12	1	2
BAR	11	47	9	14	7	0	1	2	45	0	2
7.2	8	45	5	10	5	0	1	4	18	1	1
8.4	9	55	8	7	7	0	1	4	18	1	1
MT	18	39	11	4	14	0	1	9	2	1	1
SE	13	20	24	4	16	0	0	11	4	1	1
PM	10	52	8	3	7	0	0	3	1	0	8
East											

AND: andalusite; GAR: garnet; HOR: hornblende; KYA: kyanite; MON: monazite; RAB: rutile,
anatase, brookite; SIL: sillimanite; SPH: titanite; STA: staurolite; TOU: tourmaline; ZIR: zircon

titaniferous minerals (RAB), and andalusite–sillimanite. The first cluster
represents samples situated in the Allier drainage basin and the second contains samples from the Loire basin (Fig. 6b). The Loire cluster is composed
of two groups, one with a S–N orientation in which zircon predominates and
another oriented SW–NE in which andalusite, sillimanite, and titaniferous
minerals dominate. If this cluster pattern is compared to the structural outline of the Pliocene paleo-valleys (Clozier and Turland, 1982) illustrated in
Figure 7, the orientation of the clusters corresponds to the former valleys of
the Loire, Somme, Allier, and Besbre rivers. This system of valleys therefore had an important role in transport and deposition of the Bourbonnais
Formation.

Conclusions

Correspondence analysis is most useful for processing large data sets such as
heavy mineral data, because it highlights variations which are not always apparent by simple examination of the raw data. Averages used to synthesize
large numbers of samples tend to mask variations in the data. With correspondence analysis, all samples and minerals can be studied simultaneously
and projected onto the same factor planes. As a result, not only intra-sample
and intra-mineral relationships can be visualized and interpreted, but also
their interrelationships.

To understand these relationships, the position of variables along the
factor axes must be studied together with the percentage of each mineral
contributing to each axis. In one example, minerals are grouped according
either to their metamorphic or granitic origin. Such clusters were known

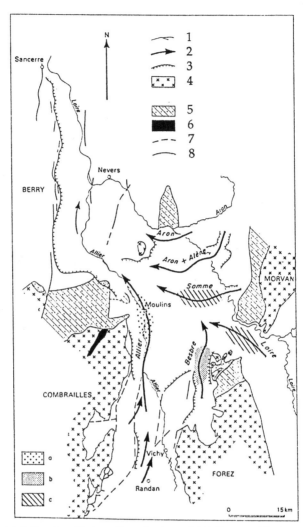

(a) Sediments with staurolite and sphene. (b) Sediments with zircons.
(c) Sediments with andalusite and sillimanite.

Figure 7: Ancient Pliocene valleys of River Allier and River Loire basin (from
Tourenq, 1989). 1: Present stream; 2: Pliocene channels (ancient valleys);
3: Limits of Plio-Pleistocene deposition; 4: Edge of crystalline basement;
5: Permo-Carboniferous; 6: Sillon Houiller Coal Measures; 7: Deep tectonic
dislocation; 8: Faults observed on surface.

prior to using correspondence analysis, but the subdivisions occurring within these groups were not known.

In the first example from the Bourbonnais Formation, the principal plane shows that the geographic position of the wells is related to the predominant heavy minerals in the samples (metamorphic minerals and zircon), indicating a paleogeographic source for the sands. The use of a third factor brings out a subdivision in the metamorphic mineral assemblages, suggesting a western origin of staurolite and an eastern origin of andalusite–sillimanite. Finally, using "supplementary" outcrop samples shows the role of ancient drainage systems in the accumulation of the Bourbonnais sands.

Acknowledgments

The authors wish to thank Dr. H. Chamley and Dr. G. de Marsily of Paris VI University, Dr. B. Beaudoin of Ecole des Mines de Paris, and Dr. A.C. Morton of the British Geological Survey for helpful criticism and suggestions that substantially improved the manuscript.

References

Ambroise, D., Gourinard, Y., Pomerol, Ch. and Tourenq, J., 1987, Stratigraphie quantitative, *in* Pomerol, Ch., (ed.), *Stratigraphie, Méthodes, Principes, Applications:* Doin, Paris, p. 235–246.

Benzécri, J.P., *et al.*, 1973, *L'analyse des Données, II. L'analyse des Correspondances:* Dunod, Paris, 619 pp.

Bolin, C., Tourenq, J. and Ambroise, D., 1982, Sédimentologie et microfossiles pyritisés du sondage de Cuise-la-Motte (Bassin de Paris): *Bull. Inf. Géol. Bassin Paris*, v. 19, p. 55–65.

Chenet, P.Y. and Teil, H., 1979, Study of some samples of hole 398D, Leg 47B, with the correspondence analysis method: *Initial Reports of the Deep Sea Drilling Project*, v. 47, p. 469–474.

Clozier, L. and Turland, M., 1982, *Notice Explicative de la Carte Géologique Dornes à 1/50 000:* BRGM, Orléans, 51 pp.

Cojan, I. and Beaudoin, B., 1986, Mise en évidence d'un contrôle paléoécologique des milieux de dépôt de bassins houillers français à partir de l'analyse des correspondances: *Bull. Soc. Nat. Elf-Aquitaine*, v. 10.2, p. 349–363.

Cojan, I., and Teil, H., 1982, Correspondence analysis used to define the paleoecological control of depositional environments of French coal basins (Westphalian and Stephanian), *in* Cubitt, J.M. and Reyment, R.A., (eds.), *Quantitative Stratigraphic Correlation:* John Wiley & Sons, Chichester, p. 249–271.

Delaune, M., Fornari, M., Herail, G., Laubacher, G. and Rouhier, M., 1989, Correlation between heavy mineral distribution and geomorphological features in the Plio-Pleistocene gold-bearing sediments in Peruvian eastern Cordillera through principal component analysis: *Bull. Soc. Geol. France*, v. 1, p. 133–144.

Faupl, P., Rohrlich, V. and Roetzel, R., 1988, Provenance of the Ottnangian sands as revealed by statistical analysis of their heavy mineral content (Austrian Molasse Zone, Upper Austria and Salzburg): *Jb. Geol. B.-A.*, Vienna, v. 131, p. 11–20.

Greenacre, M., 1984, *Theory and Applications of Correspondence Analysis:* Academic Press, Orlando, 364 pp.

Harman, H.H., 1961, *Modern Factor Analysis, 4th Ed.:* The University of Chicago Press, Chicago, 472 pp.

Imbrie, J. and van Andel, Tj. H., 1964, Vector analysis of heavy mineral data: *Geol. Soc. Amer. Bull.*, v. 75, p. 1131–1155.

Jambu, M. and Lebaux, M.O., 1983, *Cluster Analysis and Data Analysis:* North Holland, Amsterdam, 898 pp.

Lebart, L., Morineau, A. and Warwick, K.M., 1984, *Multivariate Descriptive Statistical Analysis:* John Wiley & Sons, New York, 231 pp.

Maurer, H., 1983, Sedimentpetrographische Analyse an Molassenabfolgen der Westschweitz: *Jb. Geol. B.-A.,* Vienna, v. 126, p. 23–69.

Mercier, D., Cojan, I., Beaudoin, B. and Salinas Zuniga, E., 1987, Apport des associations floristiques dans la caractérisation des paléoenvironnements sédimentaires (Bassin du Nord Pas-de-Calais): *Ann. Soc. Géol. Nord,* v. 106, p. 155–161.

Mezzadri, G. and Saccani, E., 1989, Heavy mineral distribution in Late Quaternary sediments of the southern Aegean Sea: Implication for provenance and sediment dispersal in sedimentary basins at active margins: *Jour. Sed. Petrology,* v. 59, p. 412–422.

Morton, A.C., 1985, Heavy minerals in provenance studies, *in* Zuffa, G.G., (ed.), *Provenance of Arenites:* Reidel Publ. Co., Dordrecht, p. 249–277.

Shukri, N.M., 1950, The mineralogy of some Nile sediments: *Quart. Jour. Geol. Soc. London,* v. 105 [1949], p. 511–534.

Shukri, N.M., 1951, Mineral analysis tables of some Nile sediments: *Bull. Inst. Desert d'Egypt,* v. 1, no. 2, p. 39–67.

Stattegger, K., 1986, Die Beziehungen zwischen Sediment und Hinterland Mathematisch-statistische Modelle aus Schwermineraldaten rezenter fluviatiler und fossiler Sedimente: *Jb. Geol. B.-A.,* Vienna, v. 128, p. 449–512.

Stattegger, K., 1987, Heavy minerals and provenance of sands: Modeling of lithological end members from river sands of northern Austria and from sandstones of the Austroalpine Gosau Formation (Late Cretaceous): *Jour. Sed. Petrology,* v. 57, p. 301–310.

Teil, H., 1975, Correspondence factor analysis: An outline of its method: *Math. Geol.,* v. 7, no. 1, p. 3–12.

Tourenq, J., 1986, Etude sédimentologique des alluvions de la Loire et de l'Allier, des sources au confluent. Les minéraux lourds des roches des bassins versants: *Doc. BRGM,* no. 108, BRGM, Orléans, 108 pp.

Tourenq, J., 1989, Les Sables et Argiles du Bourbonnais (Massif Central, France). Une formation fluvio-lacustre d'âge pliocène supérieur. Etude minéralogique, sédimentologique et stratigraphique: *Doc. BRGM,* no. 174, BRGM, Orléans, 333 pp.

Tourenq, J., Ambroise, D. and Rohrlich, V., 1978a, Sables et Argiles du Bourbonnais: Mise en évidence des relations entre les minéraux lourds à l'aide de l'analyse factorielle des correspondances: *Bull. Soc. Géol. France,* v. 20, p. 733–737.

Tourenq, J., Ambroise, D., Blot, G. and Turland, M., 1978b, Etude des minéraux lourds des alluvions actuelles et anciennes des bassins de l'Yonne, du Loing et de la Seine en amont de Paris. Utilisation de l'analyse factorielle des correspondances. Résultats paléogéographiques: *Bull. Inf. Géol. Bassin Paris,* v. 15, p. 25–32.

Yassin, A.A., Khalil, F.A. and El Shafie, A.G., 1984, Explanatory note to the geological map at the scale 1:2 000 000 of the Democratic Republic of the Sudan [Ed. 1981]: *Bull. Dept. Geol. and Min. Res.,* no. 35, Ministry of Energy and Mining, Khartoum, 63 pp.

6

MATHEMATICS BETWEEN SOURCE AND TRAP: UNCERTAINTY IN HYDROCARBON MIGRATION MODELING

Marek Kacewicz

Petroleum geology provides a wide spectrum of data that differs from frontier to mature areas. Data quality and quantity control which mathematical methods and techniques should be applied. In this paper two mathematical methods are shown: fuzzy-set theory and possibility theory as applied to permeability prediction and stochastic modeling of traps and leaks. Both methods are used in the modeling of hydrocarbon migration efficiency. Examples of how data uncertainty may affect final assessment of oil accumulation are presented.

Mathematical Modeling in Petroleum Geology

The complexity of petroleum geology and its importance to society stimulate research in different scientific areas including mathematical geology, which is becoming steadily more important. Armed with workstations, mainframes, and supercomputers, research laboratories in the petroleum industry investigate sophisticated mathematical techniques and develop complex mathematical models which can speed and improve exploration and lower total exploration costs. Together with classical analysis of geological, geochemical, and seismic data, mathematics provides an additional tool for basin research. The elements of petroleum systems—maturation, expulsion and primary migration, secondary migration, seals, reservoirs, and traps—may be better described by properly applied mathematical techniques.

The applicability of mathematical methods differs in frontier and mature areas and depends upon the quality and quantity of available information. Frontier areas for which data are mostly qualitative require methods which can handle imprecise and limited information easily. Fuzzy-set theory with fuzzy inference algorithms and artificial intelligence are useful approaches. Cokriging and "soft" geostatistical approaches also may be helpful. Given a limited set of parameters such as porosity or thickness, cokriging uses complementary information about seismic velocities or travel times. To obtain average rock properties, geopressures, or heat flows, seismic data can be used in mathematical inversion techniques. Mature areas, being more fully explored with hundreds or even thousands of wells and excellent seismic coverage, require spatial statistics and deterministic methods from approximation theory. Fractal theory, recently introduced to geology, is another interesting mathematical approach that has been used to describe rock properties such as fractures and porosity.

Petroleum geology requires that mathematical geologists have a good command of a wide variety of mathematical techniques. However, is this enough? It is virtually impossible to evaluate the quality of results without addressing the quality and uncertainty of data and the stability of algorithms. The quality and quantity of data can suggest which algorithms and models should be used, and which should be avoided. Even when applied to top-quality data sets, unstable algorithms can generate "garbage." We can conclude that it is not enough to know mathematical techniques and numerical algorithms; it is crucial to analyze the uncertainty in data and the numerical stability of algorithms, and to select the most optimal and appropriate methods for a given application.

This paper shows some examples of non-standard problems encountered in the modeling of secondary hydrocarbon migration in frontier areas. Assuming data are limited, we attempt to describe and account for the uncertainty in rock properties, including leakages, and to show possible migration and accumulation scenarios.

Predicting Rock Properties

Fairways of secondary hydrocarbon migration are controlled by a number of parameters. Porosity and permeability define connections or seals between source rocks and reservoirs. Porosity, because of its geometrical nature, is easier to predict than other flow parameters. There is an on-going discussion on how to find average permeability (e.g., Matheron, 1966; Le Loc'h, 1989; King, 1989). In frontier areas, the likelihood of specifying incorrect values is very high. Even slightly disturbed values, when used as parameters in the differential equations of fluid flow, may significantly change the results of modeling. The question arises: How can we avoid totally incorrect answers

Plates

Plate 1a: Oil saturation for 28.3 md basin-wide expected permeability after 4 m.y. migration.

Plate 1b: Oil saturation for 278 md basin-wide expected permeability after 4 m.y. migration.

Plate 2a: Final oil accumulations—faults are assumed to have no influence on hydrocarbon migration.

Plate 2b: Final oil accumulations—over the time of migration, faults were open to horizontal migration and after trapping, sealing faults were reactivated and leaking occurred.

Plate 3a: Final oil accumulations—over the time of migration, faults served as barriers to horizontal migration.

Plate 3b: Final oil accumulations—over the time of migration, faults served as barriers to horizontal migration but later fault activity eliminated some oil by leakage.

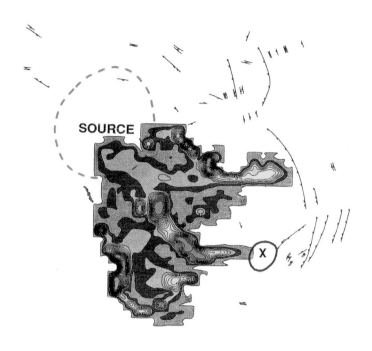

1a

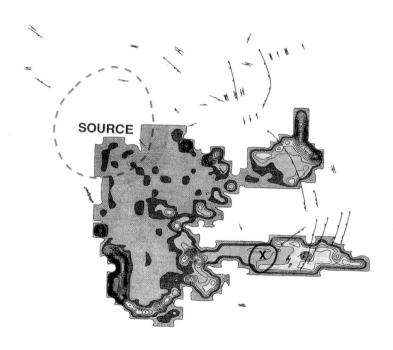

1b

2a

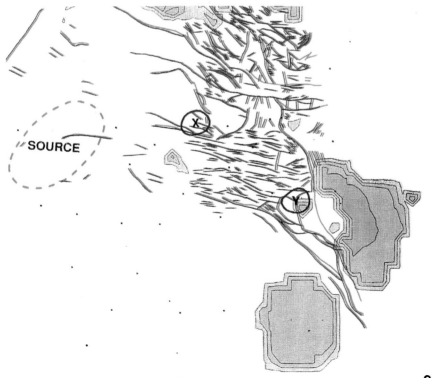

2b

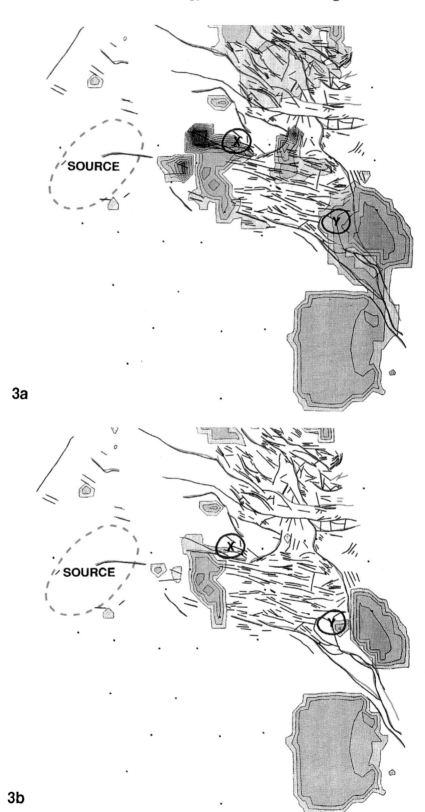

3a

3b

in basin-scale problems, where there is little or no well data? One solution is to define potential migration conduits using seismic information and geological data. Values or functions representing flow parameters are then assigned using deterministic or artificial intelligence methods. Mathematical inverse techniques can be applied to "translate" seismic information into rock parameters (Tarantola and Valette, 1982). Although these techniques are mathematically rigorous and easily programmed, final decisions about appropriate rock parameters are made arbitrarily by geologists or geophysicists. To improve the applicability of these methods, further research on uniqueness and resolution must be conducted.

Let us now express the relatively controversial view that, in some instances, precisely defined rules based on published data and our expertise can do better than deterministic and purely statistical methods. This view is especially true in areas with limited and/or poor-quality data. A methodology based on possibility theory which takes into account qualitative geological information has been discussed by Kacewicz (1989).

Uncertainty and Its Description

Describing and accounting for data uncertainty is one of the main problems encountered in modeling rock properties. Uncertainty in data commonly is represented by probability distribution functions. This corresponds to the tacit assumption, which has prevailed in science for over a hundred years, that uncertainty is equivalent to randomness (Kacewicz, 1988). Note that deterministic as well as probabilistic methods are based on the assumption that nature can be described in terms of two-level (0–1) logic, *i.e.*, a process takes place $= 1$, or does not $= 0$. As an example, a certain stratigraphic sequence will be observed with probability p versus the probability $1 - p$ that another sequence will be observed. Deterministic and probabilistic procedures do not consider circumstances in which a process or state occurs only to a certain degree. Classification of geological objects such as fossils, lithologic units, *etc.*, provides a good example. In many instances, the geological objects can be assigned to more than one class. Classic approaches consider such a situation to be an error. The natural way of thinking is to look for "sharp" limits between classes, which may cause oversimplification and limited use of information.

Prediction of rock properties in frontier areas involves similar problems. Flow simulators require assigning one permeability value to each grid cell. However, in the absence of almost any "sharp" data, permeability should be described using qualitative terms such as *low*, *medium*, or *high*, which have different meanings for different people and in different circumstances. To make these terms more formal from a mathematical point of view, a multi-level logic can be applied. Recall some basic definitions of possibility theory which have been developed from this assumption (Zadeh, 1975; Dubois and Prade, 1988; Kacewicz, 1989):

Definition 1. A fuzzy event $\mathbf{A}$ is a fuzzy subset of the elementary event $\mathbf{E}$ with a membership function $\mu_A(x)$ measurable in the Borel sense.

Definition 2. $\mathbf{A} \subseteq \mathbf{E}$ is a fuzzy event, and $p(x_1), \ldots, p(x_n)$ are probabilities for $x_1, \ldots, x_n \in \mathbf{A}$. The probability $p(\mathbf{A})$ of a fuzzy event $\mathbf{A}$ is defined as:

$$p(\mathbf{A}) = \sum_{i=1}^{n} p(x_i)\, \mu_A(x_i). \tag{1}$$

The mean value of a fuzzy event is defined as:

$$m_p(\mathbf{A}) = \frac{1}{p(\mathbf{A})} \sum_{i=1}^{n} x_i\, \mu_A(x_i)\, p(x_i). \tag{2}$$

Dispersion of a fuzzy event $\mathbf{A}$ is calculated as:

$$v_p^2(\mathbf{A}) = \frac{1}{p(\mathbf{A})} \sum_{i=1}^{n} [x_i - m_p(\mathbf{A})]^2\, \mu_A(x_i)\, p(x_i). \tag{3}$$

A membership function μ can be arbitrarily defined by an expert, based on experience and knowledge about the area considered. The probability p may be calculated, perhaps using a worldwide database, if one exists.

Fuzzy-Probabilistic Approach to Permeability Prediction

Problems with permeability (k) begin with its definition, which is based on Darcy's law:

$$U = -\frac{k}{\mu} [\mathbf{grad}\, p + \rho\, g\, \mathbf{grad}\, z], \tag{4}$$

where

$$
\begin{aligned}
U &\ - \ \text{velocity in a porous medium} \\
p &\ - \ \text{pressure of a fluid} \\
\mu &\ - \ \text{dynamic viscosity} \\
\rho &\ - \ \text{mass per unit volume} \\
g &\ - \ \text{acceleration due to gravity} \\
z &\ - \ \text{vertical coordinate.}
\end{aligned}
$$

It is obvious that discussing permeability without considering flow is meaningless. Rock samples must first be analyzed in the laboratory. The core permeabilities obtained should be "translated" into different scales, which in most instances is not a straightforward process. In frontier areas, having perhaps 3–4 wells per 40,000 sq mi, the task is even more complicated. We would like to know if the permeability of selected layers generally is very low, low, medium, high or very high. Since the states low, medium, *etc.* have different meanings for different people, they first should be defined. For example, take the fuzzy sets:

$$LPS = \left\{ \frac{1.0}{0.1}, \frac{0.8}{10}, \frac{0.6}{30}, \frac{0.4}{50}, \frac{0.1}{100} \right\} \quad \begin{array}{l} \Leftarrow \text{fuzzy membership values} \\ \Leftarrow \text{permeabilities in md} \end{array}$$

$$HPS = \left\{ \frac{0.1}{50}, \frac{0.3}{100}, \frac{0.6}{300}, \frac{0.9}{500}, \frac{1.0}{1000} \right\},$$

where LPS and HPS denote low-permeability sandstone and high-permeability sandstone. Fuzzy sets describing very low and very high permeabilities can be defined as:

$$\text{Very Low Permeability Sandstone} = LPS^m$$
$$\text{Very High Permeabiliy Sandstone} = HPS^n,$$

where

$$\mathbf{A}^p = \{x \in \mathbf{A} : \mu_{A^p}(x) = \mu_A^p(x)\}, \quad p \in \mathbf{R}. \tag{5}$$

Similarly,

$$\text{More-or-Less Low Permability Sandstone} = LPS^{1/2}$$
$$\text{More-or-Less High Permeability Sandstone} = HPS^{1/2}.$$

Medium permeability can be defined as a separate fuzzy set or as a combination of LPS and HPS:

$$\text{Medium Permeability Sandstone} = \sim LPS \wedge \sim HPS.$$

Fuzzy sets can be defined more rigorously using the following analytical function:

$$\begin{aligned} S\left(k; \alpha, \beta, \gamma\right) &= 1 & \text{for } k \leq \alpha \\ &= 1 - 2\left(\frac{k-\alpha}{\gamma-\alpha}\right)^2 & \text{for } \alpha \leq k \leq \beta \\ &= 2\left(\frac{k-\gamma}{\gamma-\alpha}\right)^2 & \text{for } \beta \leq k \leq \gamma \\ &= 0 & \text{for } k \geq \gamma. \end{aligned}$$

Figures 1, 2, and 3 show examples of LPS, VLPS $= LPS^{11}$, and $LPS^{1/2}$ fuzzy sets defined using the function S^m, where $m \in \mathbf{R}$.

Example 1: Permeability in a sandstone of the Travis Peak type.— Assume that the rock encountered in a frontier area is very similar to the Travis Peak Sandstone of East Texas because it has a similar burial history, diagenesis, *etc.* Further assume that other data from one or two wells suggest that Low Permeability Sandstone (LPS) to Very Low Permeability Sandstone (VLPS) should be expected. We can use the Travis Peak Sandstone

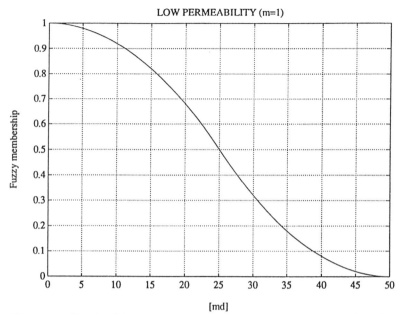

Figure 1: Graph of fuzzy set of Low Permeability Sandstone (LPS).

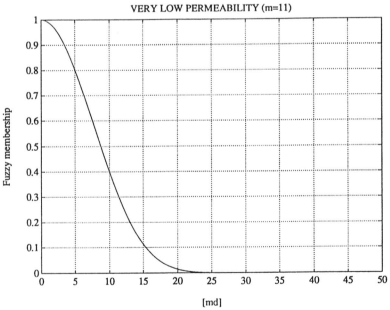

Figure 2: Graph of fuzzy set of Very Low Permeability Sandstone (VLPS).

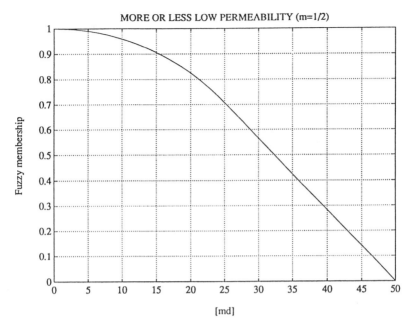

Figure 3: Graph of fuzzy set of More-or-Less Low Permeability Sandstone ($LPS^{1/2}$).

data published by Rollins, Holditch and Lee (1992, table 4), and LPS and VLPS as defined in Figures 1 and 2 to calculate an expected value and dispersion. For the LPS case the expected value is 0.83 and the dispersion is 1.45. The VLPS case has an expected value of 0.53 and dispersion of 7.66. The expected value, without taking uncertainty into account, is 1.35.

Example 2: Modeling permeability in an exploration application[1].— Differences in calculated permeability (50–100% or more) can significantly affect the final results of modeling hydrocarbon migration. For example, the oil saturations shown in Plate 1a are remarkably different from those shown in Plate 1b, not only in detail (as at prospect location X), but also in general appearance. The saturation values were obtained using a 3-dimensional, 3-phase hydrocarbon migration simulator. The only difference in input data was assuming a function for More-or-Less Medium Permeability Sandstone versus assuming a function for High Permeability Sandstone. Plate 1a–b shows oil saturations after 4 m.y. from peak expulsion for two permeability functions with calculated basin-wide expected permeabilities of 28.3 md and 278 md. Although general migration directions are similar, different migration distances are observed. The uncertainty in permeability may result in

---

[1]Because of the proprietary nature of the work, the actual location, scale, and geology of the prospects in Examples 2 and 3 cannot be released.

difficulties in interpretation. Prospect X definitely has some potential to be filled with oil, but drilling decisions must depend on which permeability scenario is considered more realistic. Additional data such as shows, oil geochemistry, and seismic profiles may be helpful in selecting the most probable answer.

Modeling Losses Between Source and Trap

Quasi-Traps and Leaks

Undetected tectonic structures in the form of barriers, traps, or faults may influence hydrocarbon migration over long distances. Some hydrocarbons are lost or entrapped and migrating hydrocarbon volumes are, at least temporarily, reduced. In frontier areas the number of undetected structures may be significant. Given limited information about tectonics, the stochastic generation of objects simulating traps and leaks may be a valuable tool. The first step is to decide how many objects to generate. An arbitrary decision may be based on statistics such as the expected number of faults or hydrocarbon fields per unit area. The selected number of objects should be generated before modeling hydrocarbon migration. The actual character of an object is inferred from the extensional or compressional character of faults where this is known, or can be generated randomly if no information is available. Two cases may be recognized: nothing is known about traps and leaks in the past or general trends are present and known (*e.g.*, known existing fault zones). Traps and leaks are considered to be "dynamic" objects. This means they change their location or character through time, corresponding to geological changes such as reactivation of faults, secondary cementation, and so forth. Statistical methods for describing the movement of traps or leaks can be similar to those used in path-planning algorithms (Reif and Sharir, 1985; Kant and Zucker, 1986; Fujimura and Samet, 1989; Kehtarnavaz and Griswold, 1990).

Cases 1 and 2 provide the most general assumptions describing movement of objects (traps or leaks) and correspond to random motion and predictable motion. It is assumed that surfaces controlling hydrocarbon migration are represented by a grid of rectangular elements of size **a**. Objects (traps or leaks) correspond to one or more grid cells.

Nothing is known about traps and leaks—random motion

Let an object O_j change its position (or character) from $(x_j(t_i), y_j(t_i))$ to $(x_j(t_{i+1}), y_j(t_{i+1}))$ (where x_k and y_k are independent random variables taking one of the values $-\mathbf{a}$, 0, $\mathbf{a}$ with equal probabilities) every Δt years. The current position of an object has been assumed or predicted. Suppose that, because of tectonic or geochemical processes, O_j renews its position, changes area, and/or changes character (trap or leak) every Δt years. An object's zone of influence on migrating hydrocarbons, which can be translated to a number of cells in a grid, is defined from the behavior of O_j in the time interval $[t_{i+1} - t_i]$. The path of an object O_j is modeled as a random vector

$$V_n = \sum_{k=1}^{n} v_k, \tag{6}$$

where $n = [(t_{i+1} - t_i)/\Delta t]$. The mean ϵ_k and variance σ_k of a random v_k are

$$\epsilon_k = E\{v_k\} = 0 \tag{7}$$

$$\sigma_k = E\{v_k \, v_k'\} = \frac{2}{3} \alpha^2 I = \sigma, \tag{8}$$

where I is the identity matrix. It follows that:

$$E\{V_n\} = 0 \tag{9}$$

$$E\{V_n \, V_n'\} = n \, \sigma. \tag{10}$$

The density of V_n approaches a normal density for reasonably large n because of the central limit theorem:

$$p_j(V_n) = \frac{1}{2\pi \left(\frac{2}{3} \, \mathbf{a}^2 n\right)} \exp\left(- \| V_n \|^2 / (4 \, \mathbf{a}^2 \, n/3)\right). \tag{11}$$

A rectangular zone of influence W_j around (x_j, y_j) is defined as:

$$P_j(V_n \in W_j) = \iint_{W_j} p_j(V_n) \, dx \, dy = H, \tag{12}$$

where H denotes the level of influence. The influence interval $(-h \, \sigma, \; h \, \sigma)$, $\sigma = (2 \, \mathbf{a}^2 \, n/3)^{1/2}$ is determined by Gaussian integration to achieve a specified level of influence H. In the first case shown in Figure 4, an object is represented by more than one cell; and in the second case, the size of an object is smaller than the grid size.

General trends are known

Positions of traps or leaks are controlled by known existing tectonic structures such as the axis of an anticline, a fault strike, or other feature. We can distinguish a stochastic approach and a deterministic approach.

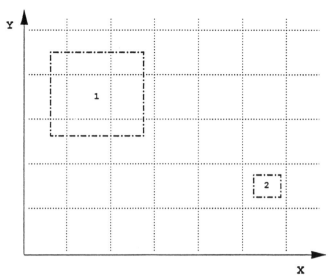

Figure 4: Grid representation of influence areas.

The stochastic approach assumes that positions $v_j(t_i)$ of an object O_j, $i = 1, 2, \ldots$ are unknown but predictable and can be determined from previous positions. $v_j(t_{i+1})$ is obtained by fitting the sequence $\{v_j(t_i),\ i = 1, 2, \ldots\}$ by an mth-order autoregressive model

$$v_j(t_i) = \sum_{k=1}^{m} \gamma(k)\, v_j(t_{i-k}) + \delta(t_i), \tag{13}$$

where $\delta(t_i)$ denotes the prediction error. An active zone of influence W_j is defined around the path $r_j(t_{i+1}) - r_j(t_i)$, where any point in W_j is located at a distance $< D$ from this path.

The deterministic approach assumes that in different time intervals the entire fault zone serves as a barrier or conductor to vertical or horizontal hydrocarbon migration. Geologically, this corresponds to compressional or extensional faulting, activation-cementation-reactivation processes, *etc.* Fault or fault-zone locations are known or can be predicted. Decisions concerning activation, cementation, reactivation, and conductivity of faults are based on reports in the literature, experience, or published formulas.

Example 3: Influence of faults on hydrocarbon migration.—An example of the influence of deterministically defined faults on long-distance secondary hydrocarbon migration is shown in Plates 2 and 3. The collision zone used in the study was determined from the grid size **a**. Green areas show final oil accumulations. Four cases are recognized:

1. No influence of faults on secondary hydrocarbon migration has been assumed, or equivalently, nothing is known about the faults (Pl. 2a).

2. Faults were open to horizontal flow over the time of migration. After trapping, faults were reactivated and leaking of the traps occurred (Pl. 2b).

3. Faults served as barriers to horizontal flow and no vertical flow is assumed (Pl. 3a).

4. Over the time of migration, faults served as barriers to horizontal migration but later fault activity eliminated some oil by leakage (Pl. 3b).

Although the general trends are similar, oil accumulations shown in Plates 2 and 3 differ significantly in detail, as they do at Prospects X and Y. The question is: Which scenario is more realistic from a geological point of view? Other geological data such as the presence of shows or seeps would be very helpful in resolving this question.

Conclusions

Differential equations and spatial statistics are not the only mathematical methods that can be applied to the modeling of secondary hydrocarbon migration. Especially in frontier areas, the lack of data or poor data quality requires us to use other, non-standard techniques. Fuzzy-set and possibility theories, together with stochastic methods, provide useful tools when data are sparse. Fuzzy-set and possibility theories are helpful in obtaining expected permeability values, taking into account our expertise and the probability (frequency) distribution functions for permeabilities in similar rocks. Undetected traps and faults and existing and known fault zones and their influence on hydrocarbon migration can be modeled using deterministic methods and the stochastic generation of traps and leaks on a grid.

The examples show how mathematical methods can help circumvent the problems of uncertainty in predicting rock permeability and describing fault activity. Multiple scenarios help in understanding general migration directions and identifying areas where further exploration should be conducted. Additional geological information always helps make hydrocarbon migration modeling more realistic.

References

Dubois, D. and Prade, H., 1988, *Possibility Theory:* Plenum Press, New York, 263 pp.

Fujimura, K. and Samet, H., 1989, A hierarchical strategy for path planning among moving obstacles: *IEEE Trans. Robotics and Automation,* v. 5, p. 61–69.

Kacewicz, M., 1988, On the description of uncertainty in geology: *De Geostatisticis,* no. 3, p. 9–10.

Kacewicz, M., 1989, On the problem of fuzzy searching for hard workability rocks in open-pit mine exploitation: *Math. Geol.,* v. 21, no. 3, p. 309–318.

Kant, K. and Zucker, S.W., 1986, Toward efficient trajectory planning: The path-velocity decomposition: *Int. Jour. Robotics Res.,* v. 5, no. 3, p. 72–89.

Kehtarnavaz, N. and Griswold, N., 1990, Establishing collision zones for obstacles moving with uncertainty: *Computer Vision, Graphics and Image Processing*, v. 49, no. 1, p. 95–103.

King, P.R., 1989, The use of renormalization for calculating effective permeability: *Transport in Porous Media*, no. 4, p. 37–58.

Le Loc'h, G., 1989, An efficient strategy for combining the permeability: Practical application on a simulated reservoir, *in* Armstrong, M., (ed.), *Geostatistics, Vol. 2:* Kluwer Academic Publ., Dordrecht, p. 557–568.

Matheron, G., 1966, Genese et signification energetique de la loi de Darcy: *Review de l'IFP et Annales des Combustibles Liquides*, v. XXI, no. 11, p. 1967–1706.

Reif, J.H. and Sharir, M., 1985, Motion planning in the presence of moving obstacles: *Proceedings IEEE Symposium on Foundations of Computer Science*, p. 144–154.

Rollins, J.B., Holditch, S.A. and Lee, W.J., 1992, Characterizing average permeability in oil and gas formations: *SPE Formation Evaluation*, v. 7, no. 1, p. 99–105.

Tarantola, A. and Valette, B., 1982, Inverse problems = Quest for information: *Jour. Geophys.*, v. 50, p. 159-170.

Zadeh, L., 1975, Calculus of fuzzy restrictions, *in* Zadeh, L., Tanaka, K. and Shimura, M., (eds.), *Fuzzy Sets and Their Applications to Cognitive and Decision Processes:* Academic Press, New York, p. 1–39.

7

RISK ANALYSIS OF PETROLEUM PROSPECTS

John W. Harbaugh
and Johannes Wendebourg

Risk analysis of an oil or gas prospect requires a probability distribution with two components, a dry-hole probability plus a distribution of oil or gas volumes if there is a discovery. While these components should be estimated objectively, risk analysis as currently practiced is mostly guesswork. Geologists assign outcome probabilities without appropriate procedures or data for objective estimation. Valid estimates require frequency data on regional exploratory drilling-success ratios, frequency distributions of oil and gas field volumes, and systematic tabulations of geological variables on a prospect-by-prospect basis. Discriminant functions can be used to analyze relationships between geological variables and hydrocarbons, leading to outcome probabilities conditional on discriminant scores. These probabilities can be incorporated in risk-analysis tables to yield risk-weighted financial forecasts. Computers are required for all procedures.

What Is Risk Analysis?

Prior to drilling a petroleum prospect, the likelihood of good outcomes must be weighed against the bad to obtain a risked financial estimate that combines all possibilities. Some oil operators simply contrast the value of discovery that is expected, versus the cost of a dry hole. A cashflow projection yields an estimate of the revenue that will be received if a discovery is made. This assumes an initial producing rate and an ultimate cumulative production for the operator's net revenue interest, and an oil price. When the stream of revenue is discounted and costs for the lease, the completed well, and operating expenses and taxes are subtracted, the net present value is obtained. If the hole is dry, its cost is readily estimated. Only two monetary estimates coupled with an intuitive guess about the likelihood of a producer versus a dry hole form the basis for a decision. A great deal of oil has been found by both independent operators and major oil companies using such simple decision systems.

Oil companies generally use more advanced methods at present. Many require their geologists to supply probability estimates for a spectrum of outcomes for each individual prospect, ranging from the probability of a dry hole through the probability of a small discovery, a medium-sized discovery, and various magnitudes of large discoveries. The probabilities are then fed into financial-analysis programs where the economic consequence of each outcome is multiplied by the probability assigned to that outcome. A bottom-line figure produced by summing the products is an *expected monetary value* or *EMV*. An EMV is a risked financial estimate suitable for making decisions but, unfortunately, obtaining valid probabilities is difficult. If the probabilities are not appropriate, the EMV is not realistic.

Probabilities can be educated guesses made by a seasoned exploration geologist who has observed many successes and failures over the years and developed an intuitive ability to estimate outcome probabilities. Unfortunately, geologists often are pressed to make prospects appear more financially attractive than warranted, because appropriately assessed prospects tend to fare poorly when ranked with competing prospects whose financial merits have been inflated.

Most major oil companies go beyond simple guesses of outcome to assess prospects and use Monte Carlo combinatorial procedures to determine probability distributions of hydrocarbon volumes. This involves multiplying the major geological factors that control volumes of producible oil or gas. The factors are expressed as probability distributions, and Monte Carlo procedures carry out the cumbersome multiplications. However, increasing the number of factors to be considered may compound the problem because the factors are so difficult to estimate. Because they are multiplied together, a single invalid factor causes the probability associated with the product to be invalid.

Some companies have developed elaborate combinatorial procedures that utilize geological, geophysical, and geochemical information. Shell Oil, for example, created a system that includes linkages between seismic and geochemical information and yields estimates of hydrocarbon charge volumes with estimates of effective regional drainage area (Sluijk and Nederlof, 1984). Shell also developed an extensive database to provide input to its system. The procedures are successful, but represent a large investment that seems to be unique in the industry.

An alternative approach is to use outcome probability estimates based on frequencies and conditional upon geology and geophysics. There is a wealth of geological, geophysical, and production information that can be used

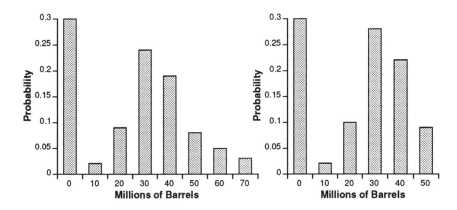

Figure 1: Outcome probability distributions for exploratory well. Left: Outcomes not constrained by area of leasehold. Right: Constrained outcomes.

to estimate conditional probabilities, and the necessary statistical tools are simple and well established. The problem is that most exploration geologists do not know how to use the tools, and personal computer programs for risk analysis have not been available.

Rules for Risk Analysis

An exploratory hole may lead to a discovery or may be dry. If it is a discovery, a spectrum of field sizes is possible. These alternative outcomes can be expressed with a single probability distribution, shown as a histogram in Figure 1. If the leasehold surrounding the prospect is smaller than areas of large fields that might be discovered, the field-volume estimates have an upper limit set by the area of the leasehold. A company is concerned only with the maximum volume of hydrocarbons that might accrue to it if a discovery is made, so the distribution may be truncated. Once an appropriate outcome probability distribution is in hand, it is readily linked with financial procedures to obtain a risked financial forecast.

A risk-analysis system requires the explorationist to provide frequencies, obtain outcome probabilities conditional on geological and geophysical properties mapped before prospects are drilled, determine conditional relationships between geological variables and hydrocarbon occurrence, map the probabilities, incorporate map error, transfer "experience" from well-explored "training areas" to frontier "target" areas, link probabilities with financial-analysis procedures, represent aversion to loss, and provide a convenient computing environment for use of these tools, displaying all steps as maps and other graphs.

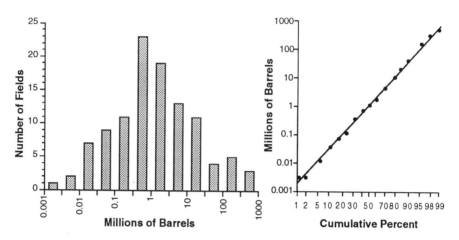

Figure 2: Big Horn field volumes. Left: Frequency distribution of estimated ultimate volumes plotted on logarithmic scale. Right: Log-probability plot.

Frequencies: Frequency data are the "ground truth" for risk analysis. Summaries of exploratory and development wells classified by outcome yield regional success ratios conditional on depth, year of completion, and producing formation and provide estimates of dry-hole probabilities. Regional frequencies of oil- and gas-field volumes provide unconditional probability distributions of hydrocarbon volumes. Cumulative production volumes for individual wells and fields can be segregated by depth, year of completion or discovery, and producing formation.

Regional field volumes can be graphed as histograms to which curves can be fitted. Figure 2 shows the frequency distribution of cumulative production in barrels of oil equivalent (BOE) for individual fields in the Big Horn basin of Wyoming and the same data in log-probability form. Both illustrate that the field volumes are lognormally distributed and suggest the form of the population of undiscovered fields in the Big Horn basin.

Probability estimates can be extracted from frequency distributions. In a region similar to the Big Horn basin, Figure 2 could provide background probability estimates. The fitted line in the log-probability plot links field volumes with probabilities on a cumulative scale, *e.g.*, if a discovery is made, the probability of finding a field containing one million barrels or more is 45%. The estimate is biased, however, because using geologic and geophysical information deliberately biases exploration; the correlation between field areas and volumes creates a second form of bias. Large fields generally form large targets and tend to be discovered before small fields. Figure 3 shows that Big Horn field sizes decline with discovery sequence, providing a quantitative measure of the effect of the combined biases.

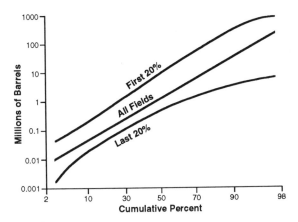

Figure 3: Log-probability plot of Big Horn field volumes. Straight line: Total population (108 fields). Upper curved line: First 20% (22 fields) discovered. Lower curved line: Last 20% (22 fields) discovered.

Conditional Outcome Probabilities: Regional frequencies are useful for background probabilities, but do not estimate probabilities for individual prospects conditional on geology. Frequency data for geological and geophysical features of prospects are seldom gathered, although they can and should be. Cross tabulations of geological properties interpreted before drilling versus outcomes obtained afterwards are needed to yield necessary conditional frequencies and conditional probabilities.

The "Pleistocene trend" of the Gulf of Mexico was extensively explored in the 1970's. Regional seismic surveys revealed structures related to salt movement that were drilled, resulting in new fields whose hydrocarbon volumes generally correlate with area of structural closure mapped seismically before drilling (Davis and Harbaugh, 1983). Probability distributions of hydrocarbon volumes conditional on area of closure as mapped before drilling can be provided by regressing discovered volume on seismic closure (Fig. 4). From the regression and its confidence intervals, we can extract probability distributions that are conditional upon acres of closure. As an example, Figure 4 shows the probability distribution for a prospect whose closure is 1000 acres. Figure 5 shows six such probability distributions, graphed in a more convenient log-probability form, that are conditional on different areas of closure and are specific for the Pleistocene trend.

Relationships between Geological Variables and Hydrocarbon Occurrence: Geological properties such as thickness, lithofacies, and structure may be only weakly related to oil occurrence when considered individually, but when considered collectively they may have a much stronger

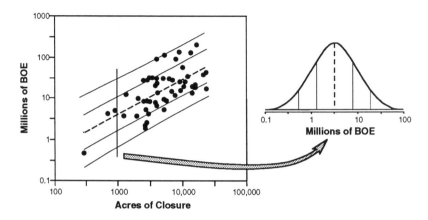

Figure 4: Pleistocene trend. Left: (Dashed line) Regression of hydrocarbon volumes in millions of barrels on acres of closure. (Solid lines) ±1 and ±2 standard deviation confidence belts around regression. Right: Lognormal distribution of field sizes corresponds to structure with 1000 acres of closure. Curve based on confidence bands around regression at left.

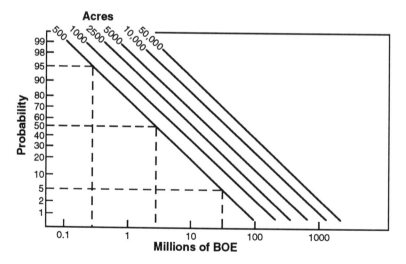

Figure 5: Family of log-probability plots of volumes of hydrocarbons corresponding to six different sizes of areas of closure. Probability estimates of 95%, 50%, and 5% on cumulative probability scale are linked with vertical and horizontal lines to corresponding hydrocarbon volumes.

relationship. Visually comparing individual maps of geological variables usually does not reveal their collective effect. Instead, we need a statistical procedure such as *discriminant function analysis* (*DFA*) that seeks out relationships that bear specifically on the presence of hydrocarbons. Discriminant analysis provides weights that transform the original variables to a single new variable called a *discriminant score*. At locations where geological variables have been mapped, their values can be inserted in the discriminant function and scores for those locations calculated and contoured. The resulting score map represents the geological variables combined in a manner that improves discrimination between favorable and unfavorable locations. For example, a discriminant function might be written

$$\text{discriminant score} = \text{coeff}_1 \times \text{thickness} + \text{coeff}_2 \times \text{shale ratio}$$
$$+ \text{coeff}_3 \times \text{structural residual}.$$

Any number of variables can be incorporated in a discriminant function. The coefficients reflect the units in which the variables are measured and also their contributions to the contrast between producing and nonproducing locations. Discriminant scores are specific for each application and generally not transferable between areas. The statistical assumptions of the methodology are described in Davis (1986) and other textbooks.

Mapping Probabilities: Although a discriminant score map is useful, it is an intermediate step toward a map of the probability of oil occurrence. A probability map links geological variables to well outcomes. Discriminant scores can be transformed to outcome probabilities using a function (Fig. 6) in which probabilities are conditional upon scores. Bayes' equation expresses the relationship between scores and conditional probabilities:

$$p(P|S) = \frac{p(S, P)}{p(S)}$$

where:

  $p(P|S)$ = conditional probability of producer, P, given knowledge of score, S
  $p(S, P)$ = joint probability (or frequency) of a producer, P, and score, S
  $p(S)$ = marginal probability (or frequency) of score, S

The function in Figure 6 was obtained by successive solutions of Bayes' equation. Discriminant scores at well locations were segregated into intervals and the numbers of wells in each outcome category in each interval were counted. For example, 13 wells have scores between 18 and 19, of which 7 are producers and 6 dry. The joint frequency of producers in this interval

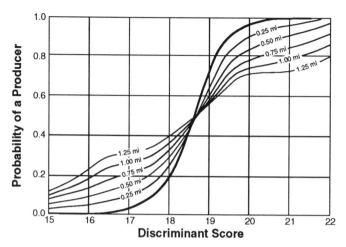

Figure 6: Heavy curve transforms discriminant scores at wells into probabilities. Lighter curves relate mapped scores to outcome probabilities at different distances from nearest well.

is 7 and the marginal frequency is 13, so the conditional probability is 7/13 = 0.54. Conditional probabilities for other intervals are obtained similarly and yield the function shown when they are plotted.

Map Error: It is impossible to produce a subsurface map without error, because it is based on the limited information available at well locations. Away from the wells, the mapped variable must be interpolated or extrapolated and is subject to error. In general, the error increases with distance from the nearest data points, although at some distance, the error may stabilize at an upper limit.

Map error may be assessed empirically, by comparing estimates made at "blind" locations with the true values, and expressing these as a distance to control as done by Harbaugh, Doveton and Davis (1977). Alternatively, the error can be estimated using geostatistical procedures. The semivariogram can be interpreted as a form of error function relating variance in estimates with distance between control and maps can be made of the expected error if the mapped property is estimated by kriging (Olea, 1975).

We can generate a function that transforms discriminant scores to probabilities and simultaneously adjusts these probabilities for map error. In Figure 6, the curves labeled with increasing distances represent specific combinations of score transforms and distance-related map error. The probability map in Figure 7 has been obtained by combining information from a map of discriminant scores and the functions in Figure 6. Figure 7 is therefore a final product that provides outcome probabilities weighted for map error over the entire area.

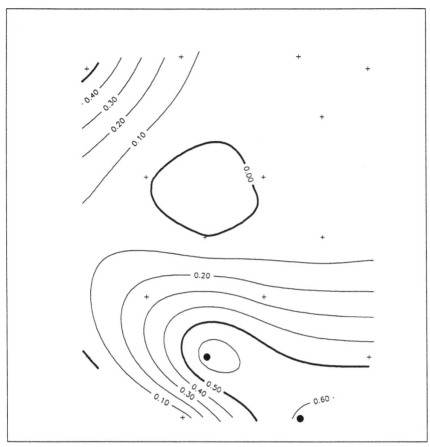

Figure 7: Contour map showing probabilities that wells will be producers, based on discriminant scores that have been transformed to probabilities. Area is 36 × 36 km. Dots are oil wells; crosses are dry holes.

Transferring Experience from Training Areas to Target Areas: Explorationists learn through personal experience and the collective experience of the industry. Experience can be quantified and formalized, and that gained in old regions can help in assessing new regions by application of relationships linking geology and oil occurrence. If this were not possible, there would be no rationale for applying geology in exploration!

Envision two areas in the same geological province. One has been maturely explored and relationships between geology and oil occurrence are relatively well understood. This can serve as a *training area*. The other, at the beginning of exploration, can be considered a *target area*. Early in the exploration of the target area it is reasonable to assume that relationships from the training area can be applied there. As the target area is progressively explored, the experience gained can supplant and eventually replace experience transferred from the training area. Figure 7 is a probability map for a target area containing only 16 wells. We are able to make the map with such limited well information because we borrowed functional relationships from a training area in the form of a discriminant function and transformed the scores to outcome probabilities and weighted them for map error. Figure 7 thus combines new information from wells within its borders and uses old relationships derived from the training area. When enough wells are drilled in the target area, its own discriminant function, as well as its own probability transform function and map error function, can be generated.

Linking Probabilities with Financial-Analysis Procedures: By itself, a probability distribution for a prospect is of scientific interest; its real usefulness is in providing a "risked" financial projection that incorporates the alternative outcomes represented by the probability distribution. Financial risking involves multiplying the financial consequences of each outcome by its associated probability and summing the products to yield the *expected monetary value* or *EMV*.

The financial elements for a prospect are conveniently expressed in a *risk-analysis table* or "*RAT*," organized so different field-size classes are columns and different financial components are rows. Rows include (a) probabilities attached to each field-size class (excluding the dry-hole probability), (b) volumes of each field-size class, (c) field areas, (d) number of producing wells, (e) average ultimate cumulative production per individual producing well, (f) average initial production per producing well, (g) present financial value of individual producing wells, (h) number of dry development wells, (i) cost of dry development wells, (j) present value, (k) readjusted probabilities incorporating the dry-hole probability, (l) risked present values, (m) exploratory dry-hole cost, (n) lease cost, (o) physical-access cost (such as road-building cost for the exploratory well), and (p) the bottom-line EMV. Such a RAT can be evaluated in a few seconds by computer, permitting the easy determination of the consequences of changes in probabilities or financial assumptions.

Aversion to Loss: The RAT can be extended to deal with risk aversion which is embodied in a *utility function* representing either a personal or corporate outlook that balances aversion to financial loss with desire for gain. A utility function transforms dollars gained or lost into arbitrary units called *utiles*. For a risk-averse individual or corporation, losses have greater negative utility than the positive utility of corresponding gains. The curvature of the function represents the degree of risk aversion. By transforming dollars to utiles, a bottom-line *expected utility value* or *EUV* can be obtained as

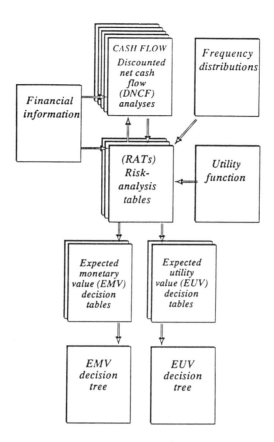

Figure 8: Diagram showing information flows in formal system for analyzing prospects.

a supplement to the EMV. Additional columns in the RAT provide entries for (q) unrisked present values in utiles, (r) risked present values in utiles, (s) negative utility of combined exploratory dry-hole cost, lease cost, and physical-access cost, and (t) the bottom-line EUV.

The central role of a RAT is to consolidate information from different sources and computer routines as outlined in Figure 8. For example, present values of individual producing wells for each field-size class are obtained by *discounted net-cash-flow analysis (DNCF)* which projects cash flows over the life of each producing well. A program for DNCF analysis therefore must be linked with the program for generating a RAT, which in turn must link with other programs that supply financial information for DNCF forecasts, probability distributions, and utility functions.

A RAT provides a bottom-line EMV or EUV for only a single action on a prospect. The action might consist of drilling the prospect with a 100% working interest (bearing all loss if the hole is dry, but receiving all production income except royalties if oil is discovered). Alternative actions might include partners sharing the working interest (reducing exposure to loss but proportionally reducing income if successful), or "farming out" the prospect to another operator who bears all expenses and exposure to loss, but returns only a fraction of the income to the prospect's originator as an overriding royalty.

A succession of RATs for alternative actions may be compared with each other by decision tables. If sequences of events such as drilling multiple prospects are analyzed, decision trees can be constructed from successions of individual decision tables.

The Computing Environment and the Need for Graphic Displays: The procedures described above require computers because they are too laborious to be done manually. To be successful, a computer implementation should be easy to use and seamlessly integrated so that information flows from one task to the next. Graphic displays are essential in most procedures, including preparation of contour maps and the display of financial results. Finally, the procedures should be available on personal computers or workstations so that geologists, managers, and financial analysts can use them conveniently. Geologists who generate prospects should be able to analyze the financial consequences of their geological assumptions and exploration managers and investors should be able to analyze how financial projections for a prospect are related to the geological and geophysical information on which the prospect is based.

Information Required for Risk Analysis

The Achilles' heel of risk analysis is the paucity of information in suitable form. It is ironic that an immense amount of potentially useful geological and production information is available, but little effort has been expended in its organization (Harbaugh, 1984). Tabulated frequencies of basic exploration and production information are needed, including outcomes of wells, cumulative production volumes for individual wells, cumulative field volumes, and field areas. Volumes should include estimated ultimate productions for individual wells and fields, organized so that subpopulations can be extracted by depth, producing formation, year of field discovery, and geological type of trap.

Tabulated frequencies of geological properties may be useful guides to hydrocarbon occurrence. These include local geological properties that can be objectively measured at the location of a prospect before drilling (such as the seismically determined area of structural closure), regional geological factors that can be assessed regionally and evaluated at prospect locations (such as regional variations in thickness of a stratigraphic interval), and regional geological properties that cannot be specifically estimated for a prospect (such as the regional thermal history).

A census should be made of past and present prospects, classified by geological features interpreted before they were drilled to yield frequencies of pre-drill interpretations. The classification should be hierarchical to permit aggregation or segregation into greater and lesser categories depending on the objectives and amount of information in the census. Production outcomes of prospects after drilling should be included to facilitate estimation of outcome probabilities conditional on geology.

Conclusions

The incentive for developing risk-assessment software is better management of risk in oil exploration which will lead to improved decision making. The advantages of using such a system are large relative to its cost. The necessary computer software for risk analysis is either available or could be developed easily because the underlying principles are simple and well understood. A challenge is to link the individual components so that information flows easily. Perhaps the greatest challenge is to organize exploration and production information in appropriate ways so that frequencies can be extracted and conditional probabilities can be estimated objectively. Geologists, managers, and financial analysts must be trained to use the data and procedures; hopefully, this is the smallest of the challenges.

Acknowledgments

We thank John C. Davis of the Kansas Geological Survey for contributions both direct and indirect to this article. John H. Doveton, also of the Kansas Geological Survey, was an earlier collaborator in our work on risk assessment and helped formulate our basic ideas. Timothy Coburn of Marathon Oil Company encouraged us to prepare computer software and instructional material for use in short courses on risk analysis at Marathon. Figures 4 and 5 are modified from Davis and Harbaugh (1983) and reproduced with permission of the American Association of Petroleum Geologists.

References

Davis, J.C., 1986, *Statistics and Data Analysis in Geology*, 2nd Ed.: John Wiley & Sons, New York, 646 pp.

Davis, J.C. and Harbaugh, J.W., 1983, Statistical appraisal of seismic prospects in Louisiana–Texas outer continental shelf: *Bulletin Am. Assoc. Petroleum Geologists*, v. 67, p. 349–358.

Harbaugh, J.W., 1984, Quantitative estimation of petroleum prospect outcome probabilities: An overview of procedures: *Marine and Petroleum Geology*, v. 1, p. 298–312.

Harbaugh, J.W., Doveton, J.H. and Davis, J.C., 1977, *Probability Methods in Oil Exploration:* Wiley-Interscience, New York, 269 pp.

Olea, R.A., 1975, *Optimum Mapping Techniques Using Regionalized Variable Theory:* Kansas Geol. Survey, Series on Spatial Analysis No. 2, 137 pp.

Sluijk, D. and Nederlof, M.H., 1894, Worldwide geological experience as a systematic basis for prospect appraisal, *in* Demaison, G. and Murris, R.J., (eds.), *Petroleum Geochemistry and Basin Evaluation:* Am. Assoc. Petroleum Geologists Memoir 35, p. 15–26.

8

CHARACTERISTIC ANALYSIS AS AN OIL EXPLORATION TOOL

H. A. F. Chaves

Characteristic analysis is well known in mineral resources appraisal and has proved useful for petroleum exploration. It also can be used to integrate geological data in sedimentary basin analysis and hydrocarbon assessment, considering geological relationships and uncertainties that result from lack of basic geological knowledge. A generalization of characteristic analysis, using fuzzy-set theory and fuzzy logic, may prove better for quantification of geologic analogues and also for description of reservoir and sedimentary facies.

Characteristic Analysis

Characteristic analysis is a discrete multivariate procedure for combining and interpreting data; Botbol (1971) originally proposed its application to geology, geochemistry, and geophysics. It has been applied mainly in the search for poorly exposed or concealed mineral deposits by exploring joint occurrences or absences of mineralogical, lithological, and structural attributes (McCammon *et al.*, 1981). It forms part of a systematic approach to resource appraisal and integration of generalized and specific geological knowledge (Chaves, 1988, 1989; Chaves and Lewis, 1989).

The technique usually requires some form of discrete sampling to be applicable—generally a spatial discretization of maps into cells or regular grids (Melo, 1988). Characteristic analysis attempts to determine the joint occurrences of various attributes that are favorable for, related to, or indicative of the occurrence of the desired phenomenon or target. In geological applications, the target usually is an economic accumulation of energy or mineral resources. Applying characteristic analysis requires the following steps: 1) the studied area is sampled using a regular square or rectangular grid of cells; 2) in each cell the favorabilities of the variables are expressed in binary or ternary form; 3) a model is chosen that indicates the cells that include the target (Sinding-Larsen *et al.*, 1979); and 4) a combined favorability map of the area is produced that points out possible new targets.

Data Transformation and Combination of Variables

The favorability of individual variables is expressed either in binary form—assigning a value of +1 to favorable and a value of 0 to unfavorable or unevaluated variables—or in ternary form if the two states represented by 0 are distinguishable—the value +1 again means favorable, the value −1 means unfavorable, and the value 0 means unevaluated. Although ternary notation is considered an improvement of the technique (McCammon *et al.*, 1983*a*), it can be used in oil exploration only in special cases because there is no method for direct detection of oil and it is risky to conclude that there is no chance for oil in specific parts of a sedimentary basin. To apply ternary notation would introduce a bias, condemning as unfavorable (−1) an area of possible oil occurrence because of unknown or unconsidered factors.

The utility of variables is established by analysis of individual maps, which is not considered part of characteristic analysis by some authors (McCammon *et al.*, 1984) but is certainly a most important phase. It is essential to quantify the geologic analogue, and this should be considered the beginning phase of characteristic analysis (Melo, 1988; Chaves and Melo, 1989). The results of characteristic analysis are completely dependent on the criteria used to choose and interpret the mapped variables for preparing the favorability map.

For data integration and information extraction, interpretation criteria can be automated in some circumstances. Calculation of local anomalies from second-derivative data or regional gradients and identification of local features on regional maps (McCammon *et al.*, 1981) are examples of interpretation criteria that can be easily automated (McCammon *et al.*, 1984).

The combination of several variables that possibly are significant for prediction of a deposit requires that the favorability of a given cell be a weighted linear combination of the transformed binary variables:

$$f(x) = a_1 + a_2 + \ldots + a_n \tag{1}$$

with weights a_i $(i = 1, 2, \ldots, n)$ and n transformed variables x_i $(i = 1, 2, \ldots, n)$. To prevent undesirable complexity, variables with weights near zero can be eliminated. In this way, characteristic analysis acts like a discriminant technique, assigning greater weights to the more discriminating variables (Chaves and Melo, 1989). Other kinds of combinations such as logical combinations (AND, OR, and NOT) of pairs of variables can be accommodated by means of specially devised truth tables (McCammon *et al.*, 1983*a*).

Model

The aim of characteristic analysis is to characterize the model—the known occurrences of hydrocarbon deposits within the target area—through relations observed among the chosen variables in the set of cells that comprises the target. It is presumed that similar relations in unknown areas will also represent favorable target areas or be clues indicating them.

The significance of each variable in the model depends upon its spatial relationships with the remaining variables, expressed by the difference between matches and mismatches among the variables taken two at a time over the subset of cells that comprise the model (McCammon *et al.*, 1983*a*, 1983*b*). For ternary-transformed variables and p cells in the model, the result varies from $+p$ in case of complete agreement for all cells (all cells have the same state, $+1$ or -1), to $-p$, when all of the cells are mismatched. Note that if one of the cells contains a zero (0) state, the comparison is disregarded.

The weights a_i in Eq. (1) are determined by solving the matrix equation

$$(\mathbf{X}' \mathbf{X})\mathbf{a} = \lambda \mathbf{a}, \tag{2}$$

where λ is the largest eigenvalue of $\mathbf{X}' \mathbf{X}$ and $\mathbf{X}$ is the $m \times n$ matrix of observations of n ternary-transformed variables (or combinations) for the m selected cells that make up the model. The a_i are the elements of the eigenvector $\mathbf{a}$ associated with λ and are scaled so that the value of $f(x)$ in Eq. (1) lies between -1 and $+1$ if a ternary form of scaling is used, or between 0 and $+1$ if a binary form is used. The solution of Eq. (2) is equivalent to maximizing

$$\sum_{i=1}^{n} \frac{\mathbf{f}' \mathbf{x}_i}{\mathbf{f}' \mathbf{f}}. \tag{3}$$

This maximized value is a measure of the overall similarity of the expression in Eq. (1) to the values for the ternary-transformed variables in the model.

Using Eq. (1), the favorability of a cell outside the model is evaluated. Values close to $+1$ indicate a close match and hence are judged highly favorable, whereas values close to -1 indicate a poor match and are judged unfavorable. Values close to zero indicate a neutral response.

To prevent a model definition that is overly restrictive because of unique features of the target, McCammon *et al.* (1983*a*) proposed a generalization that consists of adding to the k cells on the target area l cells off the target that have a high degree of match but unknown status. For each variable, the $k + l$ cells form an ordered array of $+1$'s, 0's, or -1's which can be considered as a vector. The number m of nonzero matches for each pair of variables $\mathbf{u}$ and $\mathbf{v}$ can be expressed as

$$m = \mathbf{u}' \mathbf{v}. \tag{4}$$

Assuming that the observed sequences in **u** and **v** could occur in any order, the probability that the observed number of matches is not a chance occurrence can be determined. An $n \times n$ matrix of probabilities **P** is constructed with diagonal elements equal to 1.0. **P** can now be substituted for $\mathbf{X'X}$ in Eq. (2) in the calculation of the set of weights a_i and corresponding $f(x)$ values. Those p cells of the $k + l$ cells having the highest value of $f(x)$ are selected as cells in the generalized model.

Characteristic Analysis As An Exploration Tool

The potential of characteristic analysis as an exploration tool for the oil industry was demonstrated by Melo (1988) using structural and lithostratigraphic variables in an area of known oil occurrence where the adequacy of the variables used could also be tested. The initial study was limited to publicly available data, but a later study incorporated proprietary data as well (Chaves and Melo, 1989).

Geologic Setting

An area of about 3600 km^2 of the Sergipe-Alagoas basin was chosen to test the usefulness of characteristic analysis in oil exploration. The area contains six oil fields, including Carmopolis, the largest on-shore field in Brazil. To study fracture patterns in the basement that may control fractures in the overlaying sediments, the study area was extended to the northwest to include Pre-Cambrian basement beyond the NE–SW-trending normal fault that forms the boundary of the sedimentary basin. Offshore parts of the basin to the southeast were not included in the study.

Paleozoic strata (conglomerates, sandstones, and varved shales) and Jurassic strata (red shales and sandstones) occur only around the basin margin. Cretaceous strata are more widespread and can be divided into two groups: sandstones, siltstones, and shales occur to the north and limestones and dolomitic limestones occur to the south. Tertiary sandy clays and Quaternary sands cover more than 50% of the area.

The Carmopolis oil field produces mainly from syntectonic conglomerates and is controlled by structural highs and to a lesser degree by basement fractures. Reservoirs are separated by the NW–SE-trending Carmopolis fault, with conglomerates on the northeast in the downthrown block. However, reservoirs communicate with each other and are part of the same dome structure, whose origin is related to tilting of the Sergipe-Alagoas Basin to the southeast (Lima, 1986, 1987).

The Carmopolis oil field was chosen for this study because of its importance and also because it allows testing of all variables derived from available data. Its large areal extension (47 km^2) comprises 110 cells, which is 2.5%

of the total study area (4392 cells forming a 61 × 72 matrix of 1-km × 1-km cells). Variables were defined from linear features observed on radar mosaics, fracture patterns mapped on aerial photographs, and sedimentary units, drainage, and topography measured on geological maps. The set of variables was believed to indicate possible structures in the subsurface, so areas selected by characteristic analysis are interpreted as most favorable from a structural point of view.

Application of Characteristic Analysis

Variables used in the analysis are listed below; the first nine are related to structural features, the next five are related to lithology, and the last three are logical combinations of structural and lithological variables:

1	FT	Fracture trace convergence	10	QTR	Quaternary sediments
2	FT_2	Fracture trace frequency	11	TER	Tertiary sediments
3	R_1	Presence of N47–75E lineaments	12	CRT	Cretaceous sediments
4	R_2	Presence of N48–73W lineaments	13	JUR	Jurassic sediments
5	R_3	Joint frequency of R_1 and R_2	14	PAL	Paleozoic sediments
6	R_4	Frequency of N47–75E lineaments			
7	R_5	Frequency of N48–73W lineaments	15	L_1	RIO OR QTR OR CTR
8	MFE	Morpho-structural highs	16	L_2	R_3 OR MFE
9	RIO	Drainage anomaly	17	L_3	L1 AND L .

The variables were selected and their favorability criteria set by analysis of the areal relations shown in maps of the original data. For example, fracture trace data reflect any small-scale (500 m to 3000 m) linear patterns caused by topographic features, vegetation, soil color, and segments of linear or parallel and opposite drainage. The vectorial mean of these linear features is related to joint directions measured in outcrops (Latman and Nickelsen, 1958). The measurement method used was proposed by Blanchet (1957) and developed by Chaves (1970) and is able to identify inhomogeneities in the Earth's crust (Northfleet and Bettini, 1970; Northfleet *et al.*, 1971).

The mean vector map (Fig. 1) shows areas of convergence and divergence; the former indicate relative structural highs. Melo (1988) devised a measure of the convergence in each cell (Fig. 2), which can be mapped (Fig. 3). The covergence c_j of vector mean m of each cell in the nth 3 × 3 window, as compared with z, the azimuth of the linear segment joining the center of the window with each cell center, is

$$c_j = \begin{cases} 1 & \text{when } m \subset z \pm \theta \\ 0 & \text{when } m \not\subset z \pm \theta \\ 0 & \text{when in central cell,} \end{cases}$$

where θ is a tolerance angle criterion for convergence. The convergence C for the central cell is

$$C = \sum_{j=1}^{n-1} c_j \ \ 0 \le C \le 8 .$$

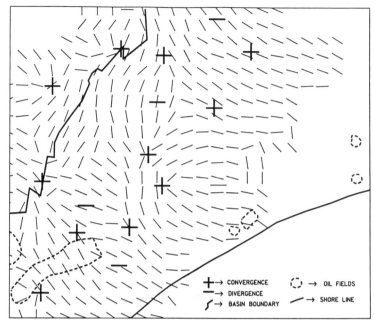

Figure 1: Vector means of fracture traces averaged within overlapping 3 × 3 cells. Study area is northeast of Carmopolis oil field, Brazil (largest dashed outline). Scale 1:500,000.

Highs indicate convergence maxima and are favorable indicators of oil traps. Highs in Figure 3 coincide with relative structural highs observed on the pre-Aptian unconformity (Fig. 4). Convergence values greater than 4 were considered favorable (+1); the rest of the area was coded unevaluated (0).

Anomalous drainage network forms have been used as indicators of geological structures. Melo (1988) used as structural high indicators the "morpho-structural" anomaly defined by Miranda (1983) as the "joint occurrence of annular, radial and asymmetric drainage forms." He also mapped radial centripetal drainage forms as possible indicators of structural lows and areas of rectilinear or almost linear drainage forms, disregarding those apparently due to consequent streams (Fig. 5). Various combinations of variables were tested, using both binary and ternary representations for data transformations (Melo, 1988; Chaves and Melo, 1989). Only Model 4b (binary) and Model 4t (ternary), which showed the most striking results, are summarized here (Table 1).

All structural variables were related to the chosen model. Lithologic variables were not especially effective, possibly because of the use of surface geologic maps; better results might be expected if maps based on subsurface data were used. All combinations of the variables were efficient in characterizing the chosen model. It is important to note that half of the target could

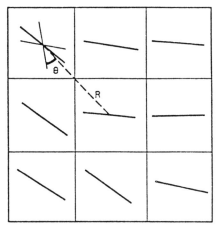

Figure 2: Calculation of mean azimuth convergence within a 3 × 3 window. R is a line joining the window center to a cell center. θ is the tolerance angle. See text for explanation.

not be efficiently characterized because the chosen variables could not discriminate the southern part of the Carmopolis oil field where oil is produced from fractures in the basement.

Ternary coding was tried (Table 1), chiefly using lithological variables (+1 for favorability, 0 for uncertainty, and −1 for absence). Binary coding (using +1 and −1) gave better results than more traditional presence–absence variables (+1, 0).

The results of regional evaluation are shown in Table 2 and in Figures 6 and 7. Both models show equivalent internal consistency; about 45% of the cells have associations above 0.8 and 95% of the cells are above 0.4. The southern part of Carmopolis oil field is above 0.4 but below 0.8. Model 4b (Fig. 6) is less restrictive than Model 4t (Fig. 7). In both models, structural variables have higher weights. A striking feature of Figure 6 is the high association south of Carmopolis field, where an extension of the field is to be tested.

Among the selected areas, at least one coincides with a seismic structure that is yet to be drilled. The most important conclusion of this research is that characteristic analysis based on properly chosen variables is applicable to oil exploration. Depending on the geologic setting, different plays will require different variables to characterize the model.

An Alternative to Characteristic Analysis

Characteristic analysis usually is defined as a multivariate technique for combining and interpreting data. An alternative way of considering characteristic

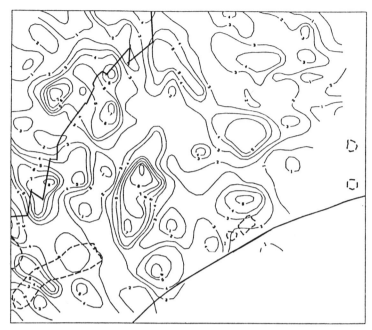

Figure 3: Isoconvergence map for fractures observed in study area northeast of Carmopolis oil field, Brazil. Scale 1:500,000.

Table 1: Variables and Weights for Model 4b Using Binary Coding and Model 4t Using Ternary Coding

Model 4b			Model 4t		
Variable*	Name	Weight	Variable*	Name	Weight
16	L_2	0.37	16	L_2	0.39
15	L_1	0.36	15	L_1	0.38
17	L_3	0.36	17	L_3	0.38
8	MFE	0.36	8	MFE	0.38
12	CRT	0.33	12	CRT	0.32
7	R_5	0.28	7	R_5	0.29
10	QTR	0.27	5	R_3	0.25
5	R_3	0.24	6	R_4	0.25
6	R_4	0.24	10	QTR	0.18
11	TER	0.19	1	FT	0.16
1	FT	0.16	2	FT_2	0.15
2	FT_2	0.14	9	RIO	0.14
9	RIO	0.13			

*Variables not listed had weights near 0.0.

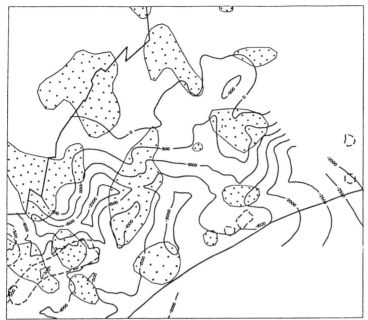

Figure 4: Structural contour map of study area drawn on pre-Aptian unconformity overlain with isoconvergence lines of magnitude 4 or greater (shaded areas). Scale 1:500,000.

analysis is by its product; that is, the final favorability map which reflects the joint occurrence of the various attributes that characterize a target. In this view, a target T is the set of cells that define a fuzzy set

$$T = \{z, f_T(z)\}, \; z \in Z, \tag{5}$$

where Z is the object space of all possible occurrences of the phenomenon and $f_T(z)$ is the degree of pertinence of z in T.

Under this definition, the pertinence function is not limited to the linear model in Eq. (1) with weights defined by the product matrix $(\mathbf{X}' \, \mathbf{X})$ or by the probability matrix $\mathbf{P}$ in Eq. (4); there are other possibilities such as the entropy function S (Dowds, 1961, 1964, 1969). This defines a measure of information:

$$S = k \sum_{c=1}^{\infty} q_e \, \log q_e, \tag{6}$$

where e are the various states and q_e are the number of choices of possible events.

By adopting a fuzzy-set definition for the model description, Boolean algebra is not required and fuzzy logic can be used instead. It is possible to express the favorability of an individual variable using fuzzy sets, opening

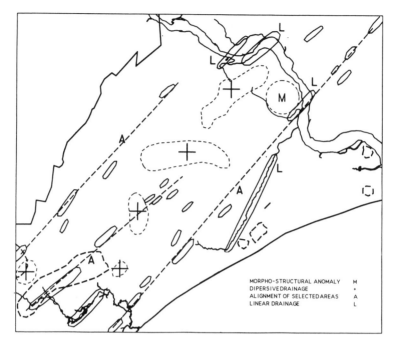

Figure 5: Map of favorable locations for occurrence of petroleum in study area, selected by drainage analysis. Scale 1:500,000.

Table 2: Results of Regional Evaluation, Model 4b (Binary)
and Model 4t (Ternary)

| | | Model 4b | | | | Model 4t | |
| | | Cells | | | | Cells | |
Class	Limit	Model	Regional	Class	Limit	Model	Regional
1	0.1	0	6.1	1	0.1	0	26.7
2	0.2	0	17.4	2	0.2	0	11.4
3	0.3	0	10.3	3	0.3	1.8	13.9
4	0.4	0.9	11.5	4	0.4	0	6.8
5	0.5	2.7	11.2	5	0.5	0	11.1
6	0.6	17.3	13.8	6	0.6	2.7	11.8
7	0.7	33.6	12.6	7	0.7	20.0	8.4
8	0.8	44.5	11.1	8	0.8	30.9	6.6
9	1.0	44.5	6.0	9	0.9	44.5	3.4

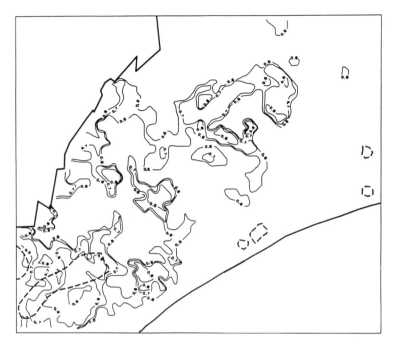

Figure 6: Model 4b favorability map. Note favorability in vicinity of Carmopolis field (largest dashed outline) extends south of field. Scale 1:500,000.

new possibilities for interpretation of variables while expressing the interpretation in a formal way.

Using variable FT_2 (fracture trace convergence), Melo (1988) determined favorable areas by comparing its map with structure on the pre-Aptian unconformity, assigning +1 to cells with convergenge values ≥ 4 and 0 to others. Under a fuzzy-set definition, the convergence values, varying from 0 to 8, could be used directly and rescaled from 0 to 1.

Eq. (1) can be generalized to

$$\mathbf{T} = \mathbf{V}_1 \cup \mathbf{V}_2 \cup \ldots \cup \mathbf{V}_n, \tag{7}$$

where $\mathbf{V}_i$ are favorability maps of individual variables defined as

$$\mathbf{V} = \{x, f\,v(x)\},\ x \in X. \tag{8}$$

$f\,v(x)$ represents the possibility that x belongs to X, the favorable area for the variable.

Applications of characteristic analysis are not restricted to regional or areal studies, and can be used to characterize reservoirs, describe oil pools, or any other set of discrete observations. The method can be applied to a reservoir as a whole, or to a reservoir discretized into cells, suggesting

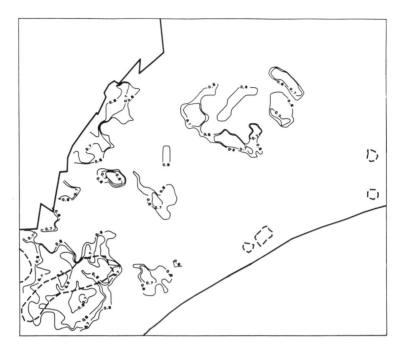

Figure 7: Model 4t favorability map of study area. Indicated favorabilities are more conservative than those in Figure 6. Scale 1:500,000.

interesting applications in reservoir simulation. Using fuzzy numbers to describe petrological and petrophysical properties of the reservoir (Romeu, 1986), characteristic analysis is well suited to describe sedimentary facies and other attributes.

There is an intuitive similarity between the concepts of pertinence functions of fuzzy sets and probability; both are used to describe problems that have high degrees of uncertainty because of random factors or incomplete knowledge. The pertinence function is subjective and can be related to subjective probability as defined by Bernoulli: "an expression of the belief someone has of an uncertain event" (Kandel and Byat, 1978). Whenever there is no appropriate pertinence function to describe the model, a hypothetical model can be used, as proposed by Sinding-Larsen *et al.* (1979).

Conclusion

Resource assessment requires integration of data including subjective knowledge and even personal feelings. Such subjective criteria are difficult to state clearly, let alone manipulate in a consistent manner. Characteristic analysis provides a way of integrating these data for sedimentary basin analysis and petroleum assessment. It produces a positive indication of areas worthy of more detailed exploration. Fruitful research issues include generalization of characteristic analysis using fuzzy-set theory and fuzzy logic and the use of other possibility functions from areas such as information theory. The application of the methodology to reservoir and sedimentary facies description may also prove to be worthwhile.

Acknowledgments

Ideas expressed here were discussed with colleagues Mary Effie Lewis, Hilton P. de Almeida, Jorge Della Favera, and Jose A. Melo. All mistakes, however, are the author's responsibility. The application of the computer program NCHARAN to petroleum exploration was conducted under the author's direction by J.A. Melo for fulfillment of his M.Sc. degree in Economic Geology at the Department of Geology of the Federal University of Rio de Janeiro. Work was supported by a scholarship and conducted at the Geoinformatic Laboratory of PETROBRAS Research Center (CENPES). Figures were produced with the assistance of Ronaldo Pereira de Oliveira using the Center's CAD system.

References

Blanchet, P.H., 1957, Development of fracture analysis as an exploration method: *Bulletin Am. Assoc. Petroleum Geologists*, v. 41, no. 8, p. 1748–1759.

Botbol, J.M., 1971, An application of characteristic analysis to mineral exploration: *Decision-Making in the Mineral Industry*, Canadian Inst. Mining and Metallurgy, Spec. Vol. 12, p. 92–99.

Chaves, H.A.F., 1970, *Appreciation of Directional Data Analysis:* Unpub. internal report [in Portuguese], PETROBRAS. RPBA. DIREX., Salvador, Bahia, 87 pp.

Chaves, H.A.F., 1988, *Characteristic Analysis Applied to Petroleum Assessment of Basins:* ILP Research Conference on Advanced Data Integration in Mineral and Energy Resources Studies, Sotogrande, Spain, 1988, 28 pp. [preprint].

Chaves, H.A.F., 1989, Generalization of characteristic analysis as a tool for petroleum assessment of basins, *in* ILP Research Conference on Advanced Data Integration in Mineral and Energy Resources Studies (Sotogrande, Spain, 1988), *Proceedings:* U.S. Geological Survey, Special Publication [in press].

Chaves, H.A.F. and Lewis, M.E., 1989, From data gathering to resources assessment: A holistic view for petroleum geology, *in* ILP Research Conference on Advanced Data Integration in Mineral and Energy Resources Studies (Sotogrande, Spain, 1988), *Proceedings:* U.S. Geological Survey, Special Publication [in press].

Chaves, H.A.F. and Melo, J.A., 1989, *Characteristic Analysis (CHARAN) Applied to Oil Exploration:* 28th International Geological Congress, Washington, D.C., Session M2: Quantitative Methods in Regional Resources Assessment [Abstract].

Dowds, J.P., 1961, Mathematical probability is an oil research tool: *World Oil*, v. 153, no. 4, p. 99–106.

Dowds, J.P., 1964, Application of information theory in establishing oil field trends: *Computers in the Mineral Industries, Part 2*, Stanford Univ. Publs. Geol. Sci., v. 9, no. 2, p. 577–610.

Dowds, J.P., 1969, Oil rocks. Information theory, Markov chains, entropy: *Colorado Sch. Mines Quart.*, v. 64, no. 3, p. 275–293.

Kandel, A. and Byat, W.J., 1978, Fuzzy sets, fuzzy algebra, and fuzzy statistics: *Proc. IEEE*, v. 66, no. 12, p. 1619–37.

Latman, L.H. and Nickelsen, R.P., 1958, Photogeologic fracture-trace mapping in Appalachian Plateau: *Bulletin Am. Assoc. Petroleum Geologists*, v. 42, no. 9, p. 2238–2245.

Lima, C.C., 1986, Structural controls on the porosity in reservoirs of the basement of Aracaju High and their implications for sedimentary reservoirs of the Sergipe-Alagoas Basin [in Portuguese]: *Boletim Tecnico da PETROBRAS*, v. 29, no. 1, p. 7–25.

Lima, C.C., 1987, *Post-Rift Structures of Sergipe Portion of the Sergipe-Alagoas Basin: Role of Tilting and Basement Discontinuities:* Unpublished M.Sc. thesis [in Portuguese], Universidade Federal de Ouro Preto, Brasil, 237 pp.

McCammon, R.B., Botbol, J.M., Sinding-Larsen, R. and Bowen, R.W., 1983a, Characteristic Analysis—1981: Final program and a possible discovery: *Math. Geol.*, v. 15, no. 1, p. 59–83.

McCammon, R.B., Botbol, J.M., McCarthy, J.H., Jr. and Gott, G.H., 1983b, Characteristic analysis applied to multiple geochemical anomalies over a concealed porphyry copper prospect, Rowe Canyon, Nevada: *AIME Trans.*, v. 272, p. 1998–2002.

McCammon, R.B., Botbol, J.M., Sinding-Larsen, R. and Bowen, R.W., 1984, The New CHARacteristic ANalysis (NCHARAN) program: *USGS Bulletin 1621*, U.S. Geological Survey, 27 pp.

Melo, J.A., 1988, *Application of characteristic analysis to hydrocarbon exploration:* Unpublished M.Sc. Thesis [in Portuguese], Universidade Federal do Rio de Janeiro, 215 pp.

Miranda, F.P., 1983, Remote Sensing in Hydrocarbon Exploration in Amazonas Basin, Brazil [in Portuguese]: *Boletim Tecnico da PETROBRAS*, v. 24, no. 4, p. 268–291.

Northfleet, A. and Bettini, C., 1970, *Fracture Analysis in the area of Miranga-Aracas, Oriental Reconcavo Basin, Bahia, Brasil:* Unpub. internal report [in Portuguese], PETROBRAS. RPBA. DIREX., Salvador, Bahia, 180 pp.

Northfleet, A., Bettini, C. and Chaves, H.A.F., 1971, Geomathematical application to petroleum prospection: Fracture analysis by orthogonal polynomials [in Portuguese]: *Congresso Brasileiro de Geologia, 23. Anais.*, v. 2, p. 61–70.

Romeu, R.K., 1986, *Logical design of a data bank for petroleum reservoirs—An application of fuzzy numbers:* Unpublished M.Sc. thesis [in Portuguese], Escola de Minas de Ouro Preto, Brazil, 120 pp.

Sinding-Larsen, R., Botbol, J.M. and McCammon, R.B., 1979, Use of weighted characteristic analysis as a tool in resource assessment, *in* Advisory Group Meeting on Evaluation of Uranium, *Proceedings:* International Atomic Energy Agency, p. 275–285.

9

INFORMATION MANAGEMENT AND MAPPING SYSTEM FOR SUBSURFACE STRATIGRAPHIC ANALYSIS

D. Gill, B. Vardi, M. Levinger, A. Toister
and A. Flexer

An information management and mapping system combining a series of interactive computer programs for stratigraphic, lithofacies, paleogeographic, and structural analysis interfaced with a comprehensive database on subsurface geology produces contour maps of quantitative variables including structure maps, isopach maps, and maps of lithofacies parameters; detailed lithologic and stratigraphic logs; and printouts of lithofacies parameters for all levels of the lithostratigraphic subdivision. Users communicate by means of simple, on-screen, menu-driven dialogues controlled by FORTRAN programs. The system runs on DEC/MicroVAX II computers operating under VMS.

The System

This information management and mapping system for subsurface stratigraphic analysis is an integration of a comprehensive database on the subsurface geology of Israel and a series of computer programs for stratigraphic, lithofacies, paleogeographic, and structural analysis. Development of the system, referred to as "ATLAS–RELIANT," was sponsored by OEIL [Israel Oil Exploration (Investment) Ltd.]. The system serves primarily as a storage and retrieval facility for information on the subsurface geology of Israel. Users can obtain printouts of lithologic and stratigraphic logs, contour maps, and value maps.

The system originally was developed to run on a CDC machine under the NOS/BE operating system. Later OEIL expanded the database to include many additional items of information [inventory of cores and petrophysical logs, results of production tests, results of petrophysical analyses, geochemical analyses of recovered fluids (water samples and hydrocarbons), and results of quantitative analyses of petrophysical logs] and the system was modified to run on DEC/MicroVAX II computers under the VMS operating system (Shertok, 1969). Among other things, the ATLAS–RELIANT system was instrumental in the regional stratigraphic analysis of the subsurface geology of Israel performed by OEIL during 1968–1988 (OEIL, 1966; Cohen *et al.*, 1990).

The Database and Its Contents

The database, dubbed "ATLAS," is about 16 MB in size and contains information on 320 petroleum exploration and development boreholes, 50 deep water wells, and 100 columnar sections of outcrops. For each borehole or outcrop the database contains identification and location information, well completion data, continuous lithologic log (condensed lithologic descriptions using 20 major rock types), lithostratigraphic and chronostratigraphic boundaries, petrophysical log markers, oil shows, porosity types and estimated amounts, nature of stratigraphic contacts, source and reliability of stratigraphic boundaries, estimates of the thickness of missing section due to erosion or insufficient penetration, fault intersections, and aggregated lithologic data for all the hierarchical levels of the lithostratigraphic subdivisions (which constitute raw data for computing lithofacies parameters). ATLAS was constructed with the SIR [Scientific Information Retrieval Inc. (Robinson *et al.*, 1979)] database management system, SIR/DBMS, and its interactive direct data entry and editing facility, SIR/FORMS. A detailed review of the database structure and its information content is presented in Flexer *et al.* (1981*a*, 1981*b*).

Interactive Programs for Stratigraphic Analysis

ATLAS interfaces with a system of interactive programs for general stratigraphic analysis referred to as "RELIANT" (REgional LIthofacies Analysis Tool—or Toy). The system is user-friendly and does not require any previous experience in automated data processing. User/machine communication is by means of a simple, on-screen, menu-driven dialogue. The system prompts the user with explicit instructions and for most responses provides a menu of options. It contains adequate features for error recovery so that job termination due to user mistakes is rare. The dialogue is controlled by FORTRAN programs. Each dialogue activates specific procedures written in the programming language of the operating system. These procedures link RELIANT programs (written in the SIR command language or in FOR-TRAN) with the correct files, and initiate batch jobs which carry out the user's requests.

The user can obtain data printouts, sorted and formatted to his specifications, for most types of data. These include detailed lithologic and stratigraphic logs and printouts of lithofacies parameters for all levels of the lithostratigraphic subdivision. Two types of maps can be prepared: (1) contour maps of quantitative variables, and (2) value maps, which are base-map postings of qualitative variables. Contour maps include structure maps, isopach maps, and maps of lithofacies parameters such as the percentage, ratio, and number of occurrences of certain lithic components. Value maps show (1) unit occurrence at a designated depth, (2) the identity of lithostratigraphic units immediately above ("worm's-eye" view) or below ("bird's-eye" view) a designated horizon or depth, (3) facies classes according to two three-component facies classification schemes, and (4) oil shows. Maps are prepared by RELIANT using the SURFACE II graphics package developed at the Kansas Geological Survey (Sampson, 1978), which is integrated into the system.

Correlating Stratigraphic Interpretations

In order to prepare country-wide maps, it was necessary to establish the correlation between lithostratigraphic subdivisions introduced by different authors and/or employed in different parts of the country. The unified scheme prepared for this project is based on a compilation of data from numerous sources [see Flexer *et al.* (1981a) and references therein]. Furthermore, the detailed subdivision of each case into rock- and time-stratigraphic units had to be worked out before it could be admitted into the database. It is not uncommon to find that authors differ in their stratigraphic interpretations of certain cases. Therefore, the system was designed to accommodate more than one stratigraphic interpretation per case. However, because geologists may differ in their interpretation of regional stratigraphic correlations, this may be insufficient to suit everybody. In order to allow for a multiplicity of regional stratigraphic schemes, substantial modifications must be made in the database and the supporting software.

Developing the System

The various stages and activities involved in designing and implementing the ATLAS–RELIANT system are summarized schematically in Figure 1. The process can be described in terms of three stages, in which three entities take part. The compartments in this matrix consist of certain activities,

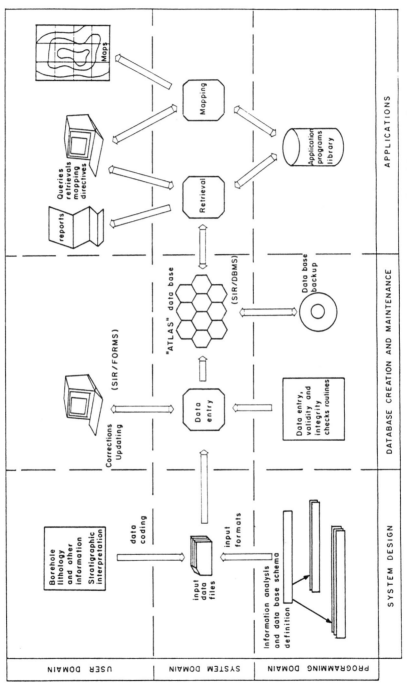

Figure 1: Diagrammatic layout of the design, development, and use of the ATLAS-RELIANT system.

software components, or a combination of the two. The main responsibilities of the geologist are to compile the country-wide stratigraphic correlation table discussed above, define the stratigraphic subdivision of the cases, and verify the completeness and correctness of all data items in the database. At the design stage, the main tasks of the programmer and system analyst are to define the architecture and schema of the database which will best implement all the anticipated applications, to devise automatic input error detection features, and to prepare the application software. Between the programmer and the user is a stratum of software components through which the user communicates with the system.

Map Preparation

The ATLAS–RELIANT system is, first and foremost, a depository for geological information on the subsurface geology of Israel. Its principal utility is for storage and retrieval of this information for any task. The preparation of maps involves three principal steps (Fig. 2). The first consists of selective retrieval of data from the database into a designated working file. The retrieval program can accommodate requests constrained by any combination and limiting values of database variables, with practically no limit on complexity. Thus, data can be screened by very specific constraints. The user defines the map type, horizon or interval of interest, and, in the case of lithofacies maps, the relevant lithic components to be considered. At this stage the working file will contain all the cases that meet the specified constraints. The next program performs the posting or contouring within the geographic boundaries defined by the user (by reference to predetermined regions or by coordinates). Optional parameters include the map scale, the density of the interpolation grid, and the amount of smoothing of the contour lines. At this stage the user also has the option to delete specific boreholes from the map. The map may be previewed on a graphics terminal and changes can be made by returning to the previous step. After validation the map can be drawn on paper by a plotter. A detailed user's manual of the system is presented in Gill *et al.* (1982).

Acknowledgments

The authors extend grateful thanks to Y. Langotsky, former director of OEIL, for initiating and supporting the ATLAS–RELIANT system project; to N. Shafran, A. Yelin-Dror, and A. Livnat of Tel Aviv University and to Y. Weiler, L. Fleisher, and E. Shertok of OEIL for their efforts and assistance throughout the project. We also thank A. Peer for drafting the illustrations and I. Perath for helpful editorial suggestions.

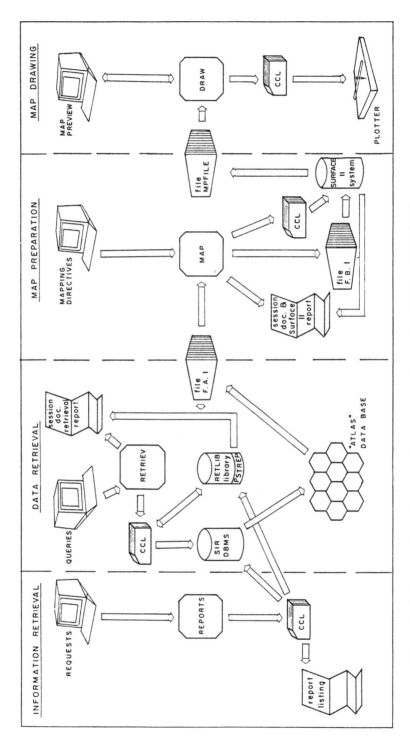

Figure 2: Schematic flow diagram of map preparation using ATLAS-RELIANT system.

References

Cohen, Z., Kaftzan, V. and Flexer, A., 1990, The tectonic mosaic of the southern Levant: Implications for hydrocarbon prospects: *Jour. Petroleum Geol.*, v. 13, p. 437–462.

Flexer, A., Gill, D., Livnat, A., Tamir, N. and Toister, A., 1981*a*, *ATLAS Project, Part A— Explanatory Notes:* Oil Exploration (Investment) Ltd. and Ramot, Tel Aviv University Applied Research and Industrial Development Adm., Report No. 1/81, 43 pp. [in Hebrew].

Flexer, A., Gill, D., Livnat, A., Toister, A., Shafran, N. and Tamir, N., 1981*b*, Computerized information system in support of oil and gas exploration in Israel: *Israel Geol. Soc. Abstracts*, Symposium on Oil Exploration in Israel, p. 16–20.

Gill, D., Vardi, B. and Ataya, H., 1982, *"Atlas" Project—User Guide to Mapping Software System: Seminar Notes, Atlas Users' Seminar:* Israel Geol. Survey, Unpub. report, 13 pp. [in Hebrew].

OEIL, 1988, *Hydrocarbon Potential of Israel:* Oil Exploration (Investment) Ltd., Unpub. report, 148 pp.

Robinson, B.N., Anderson, G.D., Cohen, E. and Gazdzik, W.F., 1979, *Scientific Information Retrieval User's Manual:* SIR Inc., Evanston, Illinois.

Sampson, R.J., 1978, *SURFACE II Graphics System:* Kansas Geol. Survey, Series on Spatial Analysis No. 1, 240 pp.

Shertok, E., 1989, *ATLAS CPS1 Users' Manual:* Oil Exploration (Investment) Ltd., Report 89/3, 18 pp.

10

AUTOMATED CORRELATION BASED ON MARKOV ANALYSIS OF VERTICAL SUCCESSIONS AND WALTHER'S LAW

D.R. Collins and J.H. Doveton

Walther's Law of Facies (1894) states that facies overlying one another comformably were formed in geographically contiguous environments. This vertical-lateral linkage is the basis for our automated method of stratigraphic correlation. The probabilities of vertical adjacency of different lithologies are estimated by embedded Markov chain analysis of sequences to be correlated. These probabilities are transformed to dissimilarities and used as elements within a dynamic programming sequence comparison. Trajectory tracking of cumulative thicknesses between the two sequences provides an auxiliary criterion to incorporate factors of sedimentation rate and compaction.

Establishing Lithostratigraphic Equivalence

Stratigraphic correlation is simultaneously simple and complex. The operation is fundamentally one of pattern recognition, whose principles can be grasped easily by any geology student. One source of complexity is caused by the fact that most successions are composed of a relatively small number of distinctive rock types. Within each succession, they are ordered as a linear chain in which loosely repetitive sequences are often perceived as "cycles" or "rhythms." As a result, the correlation between two adjacent successions may be ambiguous, so that several competing alternatives may be equally valid candidates for the "true" correlation.

The situation is made still more dismal by the knowledge that erosional events may have removed entire stratigraphic segments and that periods of non-deposition may have caused gaps. In the opinion of Ager (1973), the gap is more important than the record. Even if a "complete" lithology record were available, it is unlikely that the successions in two separate locations would be identical. Lateral facies changes result in differences of lithology within correlative intervals.

Equivalence or "similarity" of rock type is not the only criterion used in correlation. Thicknesses are a secondary source of information for correlation decisions. Similarity in thickness of equivalent lithologies between successions often implies a greater likelihood of their correlation. However, exceptions to this rule commonly are observed in the lateral thinnings and thickenings caused by both lateral facies changes and differential compaction.

The simpler aspects of correlation suggest that practical automated correlation procedures are both feasible and desirable. Even if programmed decisions cannot be characterized absolutely as "objective," they can at least be made consistent. Furthermore, the application of dynamic programming methods, which are the core of recent correlation algorithms, allows the simultaneous consideration of all possible correlation alternatives. This is a difficult feat for the human correlator to achieve. In this paper, we introduce a quantitative correlation method which has two principal characteristics. First, the decision procedures are couched in terms of probability, which accommodates the inevitable sequence disruptions and ambiguities of natural succession. Second, the method draws on the principle of Walther's Law to provide a link between vertical succession and lateral correlation.

Johannes Walther was one of the earliest proponents of the facies concept and used the term *Faziesbezirk* to describe a set of genetically related facies ordered in a comfortable vertical sequence. He also identified erosion, deposition, and equilibrium as modifiers of his "faziesbezirk" unit. Walther (1894) recognized that facies that succeed one another conformably must have been deposited in adjacent environments. This concept is now known as "Walther's Law" and is summarized by Selley (1976) as "a conformable vertical sequence of facies... generated by a lateral sequence of environments."

Walther's Law has important implications for correlation. Information concerning the pattern of vertical ordering of sedimentary lithologies is the key to the recognition of lateral equivalents between neighboring successions. However, even a casual observation of a vertical succession reveals that patterns in vertical ordering are by no means deterministic, unless a forbiddingly complex history of unconformities and non-depositional events is invoked. Instead, a stochastic model is appropriate which accounts for vertical adjacencies in terms of probability. An embedded Markov chain is the simplest stochastic model for the description and analysis of ordering patterns in rock successions and is described by Doveton (1971).

A contingency table of the frequencies of observed vertical transitions between lithology types within two successions to be correlated may be converted into a transition probability matrix. These probabilities express not only the likelihood of vertical adjacency, but function as estimates of their likelihood of being laterally adjacent facies between successions. It is important to realize that Markovian analysis does not impose hypothetical ideal patterns of ordering commonly contained in notions of "cycles," "rhythms," or similar terms. If, for example, the successions are riddled with unconformities which have partially randomized ordering, this situation will be reflected faithfully by the transition probabilities. However, loosely repetitive motifs are commonly observed and their character will be expressed in the transition matrix.

Correlation of two adjacent successions can be made by merging them into a single composite sequence whose cumulative transition probability product is the maximum possible. The optimal sequence can be found by the dynamic programming algorithm applied to correlation by Smith and Waterman (1980), Howell (1983), Waterman and Raymond (1987), Wu and Nyland (1987), Lineman *et al.* (1987), Griffiths and Bakke (1988), and others. These authors have drawn on methods with proven success in sequence comparison within linguistics, molecular chains, and birdsong characterization, which are summarized by Sankoff and Kruskal (1983).

All these correlation procedures have grappled with the same problem: How can dissimilarities between lithologies be established as systematic measures which are not arbitrary, but reflect real estimates of stratigraphic dissimilarity? We suggest that transition probabilities provide the appropriate measure since they are based on observational data rather than hypothetical external criteria. Also, their probabilistic nature accommodates the degree of uncertainty that is inherent in all correlation solutions.

Figure 1 is a schematic example of the procedure, where sequence A is correlated with sequence B. The dynamic programming framework uses the two sequences as axes, with the origin in the top left corner equated with a "null" state conventionally denoted as Φ. The algorithm seeks to find an optimum pathway through the matrix such that the cumulative transition probability product is a maximum. The search is constrained to honor stratigraphic order by moving down and to the right, and applied recursively at each cell location. At each path step, the possible transitions from immediately earlier cells are compared, and the most probable linkage selected. By accumulating a running probability product across the matrix, the optimum correlation can be found by a traceback of the maximum probability pathway. The optimum transition pathway of sequences A and B is shown together with their correlation in Figure 1. As already stated, the process can be thought of as the merger of the two sequences as a single composite in the "slotting" sense (see Gordon and Reyment, 1979). Alternatively, it can be

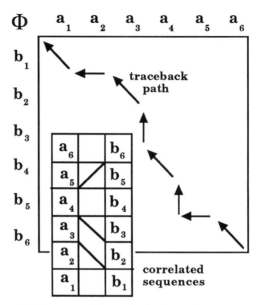

Figure 1: Schematic diagram of hypothetical correlation using dynamic programming in sequence comparison.

related to the seminal concepts of Levenshtein (1966) who defined three basic operations in sequence comparisons: matching (*e.g.*, a_3 and b_2), deletion (*e.g.*, a_2), and insertion (*e.g.*, b_3). The correlation linkages shown between the two sequences describe bounding surfaces which can be interpreted in terms of lithostratigraphic equivalence, lithofacies changes, pinchouts, and unconformities. The method is explained in the following case study, where additional concepts concerning bed thickness are also incorporated in the selection of an optimum correlation.

A Stratigraphic Correlation Case Study

Stratigraphic sequences formed from deltaic deposits are instructive material for testing any automated correlation method. Some facies, such as marine shales, constitute widespread marker horizons. By contrast, clastic facies associated with distributary channel networks are limited in areal extent and grade laterally into adjacent facies. While these sequences are not random, neither are they expressions of simplistic "layercake geology." The Pennsylvanian Coal Measures of Ayrshire, Scotland, appear to have originated in a delta-plain environment where occasional weak marine transgressions deposited localized shales containing marine or brackish water fauna (Mykura, 1967). Six distinct lithologies which are usable states for Markov and correlation analyses are recognized in borehole core records: "barren" shales

Table 1: Markov Tally/Probability Matrix for Vertical Transitions between
Lithological States Observed in the Successions of Two Neighboring
Boreholes. (Probabilities are in parentheses; key to coding of
lithologic state is given in text.)

	A	M	B	C	Y
A		4(0.27)	6(0.40)	0(0.00)	5(0.33)
M	7(0.70)		3(0.30)	0(0.00)	0(0.00)
B	5(0.31)	3(0.19)		6(0.38)	2(0.12)
C	0(0.00)	0(0.00)	6(1.00)		0(0.00)
Y	3(0.43)	3(0.43)	1(0.14)	0(0.00)	

or mudstones (A), shales with non-marine bivalves (M), siltstones (B), thin
sheet sandstones (C), thick channel sandstones (K) (not present in all sections), and rootlet horizons with or without an overlying coal (Y).

The interval between the Ell and Main coals was correlated between sequences cored in several boreholes in the Ayrshire coal field. An embedded
Markov tally matrix of lithology transitions from two neighboring boreholes
is shown in Table 1, together with the resulting transition probability matrix.
An optimum correlation can be found by melding the two separate sequences
into a composite sequence, in which the product of transition probabilities is
a maximum. The most effective method for this purpose is the dynamic programming algorithm described by Delcoigne and Hansen (1975) and applied
by later authors. Dynamic programming of the correlation problem presup-
poses the assignment of dissimilarities associated with possible matches or
gap insertions. These are accumulated additively while tracing an optimum
correlation pathway across a framework whose axes are the two sequences
to be correlated. This additive character can be accommodated by using
the criterion equivalence:

$$\text{maximum } \Pi \ p_{ij} \equiv \text{minimum } \Sigma - \log \ p_{ij}$$

where p_{ij} is the probability of transition from state i to state j.

Transition probabilities of a state to itself take unit value and hence have
a logarithmic dissimilarity of zero. Zero transition probabilities imply infinitely large negative logarithmic dissimilarity values, but were assigned an
arbitrary high number for practical programming. This convention assures
that transitions or equivalences will be precluded between states which are
never observed to be adjacent in their vertical ordering.

The optimum traceback path is shown for the correlation of the two
boreholes in Figure 2. The path is broken locally by bifurcation networks

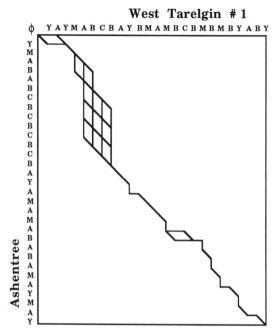

Figure 2: Dynamic programming framework and traceback path for correlation of two lithological successions, using Markov transition probabilities as path-guide criteria. The lettering convention for the lithologies is explained in the text.

which reflect repetitive sequences in which several alternative and equally valid correlations are possible. However, it is important to note that a fundamental limitation of the framework is that thickness information is not considered. Instead, equal "thicknesses" are implied for all lithologies in both successions. Thickness is therefore a useful auxiliary criterion to select the most likely correlation route within the ambiguous parts of this network. An even more compelling reason for considering thickness is that comparisons of thickness are an integral component in the decision process of a human correlator.

Some authors who used the dynamic programming method for stratigraphic sequence correlation ignored thickness altogether. Others, such as Waterman and Raymond (1987), have struggled with the definition of a dissimilarity function based on differences in thickness between potentially correlatable beds. Part of the difficulty results when dissimilarities in thickness measured in feet must be combined with lithological dissimilarities based on grain-size differences, transition probabilities, or other properties. Inevitably, "apples" are mixed with "oranges." The problem is further complicated by the fact that some lithologies show dramatic lateral changes in

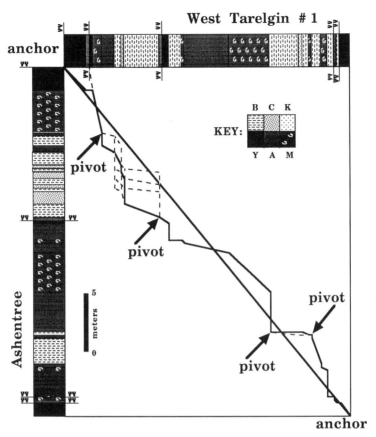

Figure 3: Dynamic programming framework of Figure 2 rescaled in thickness units with optimum traceback correlation. The lettering convention for the lithology key is explained in the text.

thickness, while the thicknesses of other lithologies remain essentially constant over long distances.

We suggest that thickness data should be treated in a different fashion. Rather than focusing on thickness differences between individual beds and their potential correlatives, attention should be directed to comparisons of cumulative thickness. If the axes of the dynamic programming framework are rescaled in units of thickness, the optimum traceback network derived from lithological state correlation now traces out an implied record of sedimentation, compaction, non-deposition, erosion, and faulting. This transformation of the raw dynamic programming framework is shown in Figure 3.

A main diagonal straight line drawn from the bases of the two successions to their tops defines a first-order model. By selecting the traceback path which has the minimum deviation from this line, continuous synchronization in sedimentation rate and compaction is assumed between the two successions. Such a model is unnecessarily simplistic and inherently limited. In a second-order model, a criterion of "smoothness" may be applied to localize synchronization. On Figure 3, local zones of ambiguity are bounded by "pivots" located at the ends of unique pathway segments. The complete network is "anchored" by the correlatable tops and bases of the two sequences. The trends of the unique pathway segments are then interpolated into the bifurcation network zones, so that the final result is the smoothest trajectory possible. The complete trajectory has intrinsic interest because it sketches out an interpretation of the relationship in sedimentation and compaction history between the two successions.

The final result of this process is shown in Figure 4, where the correlation between the two borehole sections is shown in a conventional graphic form. Notice that while many correlations are made between common lithological states, others are "implied" correlations of facies transitions, and it is suggested that others thin out and disappear between the two boreholes. The simplicity of both the algorithms and the criteria used in this approach are relatively easy to code as a computer program, which can be applied to correlating any type of stratigraphic sequence. As a more extensive example, pairwise correlations between four boreholes are shown in Figure 5.

Discussion

The analytical strategy described in this paper is novel, although its components include Walther's Law, embedded Markov chain analysis, and the dynamic programming of sequence comparisons that have been tested extensively in many geological applications. Unfortunately, the ambiguities of alternative correlations in complex successions will forever be facts of life. The present limitations and potential future improvements in the performance of this method are discussed below.

In this application, transition probability matrices were estimated from the successions to be correlated. In most instances, transition cell frequency counts will be relatively low and will yield small sample estimates of the parameters. However, short successions can be considered to be locally stationary, even though the sample counts are smaller than desirable. Finally, and importantly, it should be noted that precision in the estimates is not the goal of this procedure, but rather the correct monotonic ordering of transition weights to be used in selecting among the alternative correlation pathways.

Transition probabilities between lithological states have been used as the measure of local adjacency. Alternative measures from Markovian analysis are available. Doveton and Duff (1984) used mean first passage times for characterizing the ordering in Illinois coal measures. The mean first passage

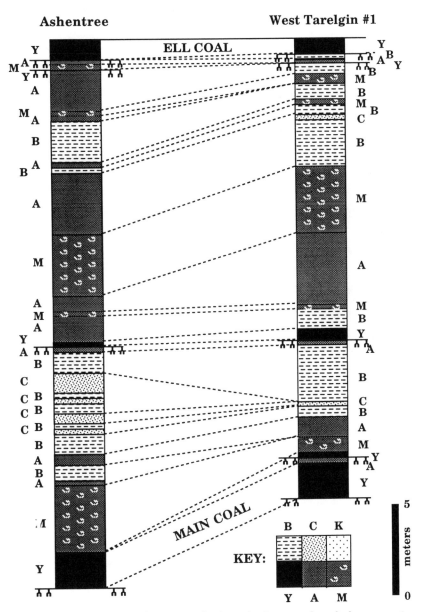

Figure 4: Result of correlation method applied to two borehole successions. The lettering convention for the lithology key is explained in the text.

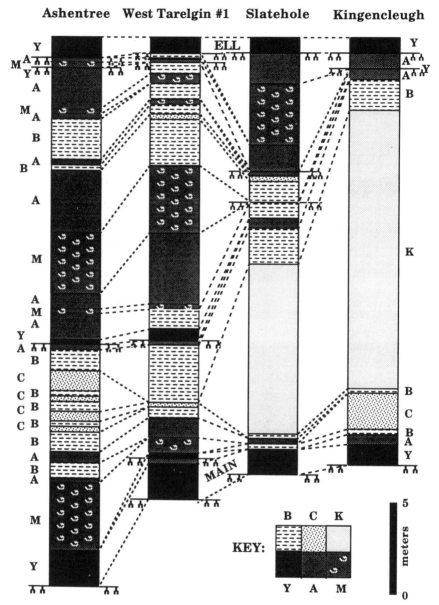

Figure 5: Correlation of four boreholes from the Ayrshire Coal Measures (Pennsylvanian, Scotland).

time matrix may easily be calculated from the Markov transition probability matrix as described by Kemeny and Snell (1960). It represents the average number of states which occur between one state and the first occurrence of a second state. The mean first passage time can be inserted into the correlation algorithm as an alternative estimate of dissimilarity between pairs of states. Although transition probabilities are more easily interpreted, the use of passage times may improve correlation performance because they are based on the entire network of transitions in the transition probability matrix.

The notion of comparing the cumulative thickness trajectory between two successions is an attractive concept, but requires some operational decisions to make it work effectively. In particular, the interpolation of a locally smooth trajectory segment between pivots can be made to honor a variety of criteria. Further experimentation is necessary to determine whether local polynomial curves, splines, or other functions provide the most reliable prediction in situations where correlations are known.

Finally, the correlation linkage that results from this method may be found locally to be incorrect for sound geological reasons which override the criteria of the method. This problem is common to all automated correlation procedures. The method allows for any degree of manual intervention. Where correlations between certain horizons within the two sequences are known, local "anchors" may be established as points that will constrain the route of the correlation pathway.

References

Ager, D.V., 1973, *The Nature of the Stratigraphical Record:* John Wiley & Sons, New York, 114 pp.

Delcoigne, A. and Hansen, P., 1975, Sequence comparison by dynamic programming: *Biometrika,* v. 62, p. 661–664.

Doveton, J.H., 1971, An application of Markov chain analysis to the Ayrshire Coal Measures succession: *Scott. J. Geol.,* v. 7, p. 11–27.

Doveton, J.H. and Duff, P. McL. D., 1984, Passage-time characteristics of Pennsylvanian sequences in Illinois, Ninth ICC Congr.: *Compte Rendu,* v. 3, p. 599–604.

Gordon, A.D. and Reyment, R.A., 1979, Slotting of borehole sequences: *Math. Geol.,* v. 11, p. 309–327.

Griffiths, C.M. and Bakke, S., 1988, Semi-automated well matching using gene-typing algorithms and a numerical lithology derived from wire-line logs: *Trans. SPWLA,* 29th Ann. Logging Symposium, Paper GG, 24 pp.

Howell, J.A., 1983, A FORTRAN 77 program for automatic stratigraphic correlation: *Computers & Geosci.,* v. 9, p. 311–327.

Kemeny, J.G. and Snell, J.L., 1960, *Finite Markov Chains:* Van Nostrand, Princeton, 210 pp.

Levenshtein, V.I., 1966, Binary codes capable of correcting deletions, insertions, and reversals: *Cybernetics and Control Theory,* v. 10, no. 8, p. 707–710.

Lineman, D.J., Mendelson, J.D. and Toksoz, M.N., 1987, Well to well log correlation using knowledge-based systems and dynamic depth warping: *Trans. SPWLA,* 28th Ann. Logging Symposium, Paper UU, 25 pp.

Mykura, W., 1967, The Upper Carboniferous rocks of south-west Ayrshire: *Bull. Geol. Surv. Gt. Brit.,* no. 26, p. 12–98.

Sankoff, D. and Kruskal, J.B. [eds.], 1983, *Time Warps, String Edits, and Macromolecules: The Theory and Practice of Sequence Comparisons:* Addison-Wesley, London, 328 pp.

Selley, R.C., 1976, *An Introduction to Sedimentology:* Academic Press, London, 408 pp.

Smith, T.F. and Waterman, M.S., 1980, New stratigraphic correlation techniques: *Jour. Geol.,* v. 88, no. 4, p. 451–457.

Walther, J., 1894, *Einleitung in die Geologie als Historische Wissenschaft, Bd. 3, Lithogenesis der Gegenwart:* G. Fischer, Jena, p. 535–1055.

Waterman, M.S. and Raymond, R., Jr., 1987, The match game: New stratigraphic correlation algorithms: *Math. Geol.,* v. 19, p. 109–127.

Wu, X. and Nyland, E., 1987, Automated stratigraphic interpretation of well log data: *Geophysics,* v. 52, no. 12, p. 1665–1676.

11

MILANKOVITCH CYCLICITY IN THE STRATIGRAPHIC RECORD— A REVIEW

S.B. Kelly and J.M. Cubitt

The Milankovitch or astronomical theory of paleoclimates relates climatic variation to the amount of solar energy available at the Earth's surface. The theory helps explain periodic, climatically related phenomena such as the Pleistocene ice ages. Identification of Milankovitch cyclicity within sediments demonstrates the influence of climate on sedimentation patterns and creates a time frame for the estimation of basin subsidence rates.

Spectral analysis of deep sea and ice cores indicates periodic climatic fluctuations during Tertiary and Quaternary times. These fluctuations are strongly cyclical with low frequencies centered at periods around 400 ka and 100 ka together with shorter periodic components of approximately 41 and 21 ka. Lower frequencies reflect eccentricity of the Earth's orbit; 41- and 21-ka components are associated with periodic changes in the tilt of the Earth's axis and the precession of the equinoxes.

Astronomically forced glacial eustasy results in distinct stratigraphic units or parasequences of widespread extent. Milankovitch band parasequences occur in both carbonate and clastic shelf systems, including cyclothemic Upper Paleozoic successions of North America.

Milankovitch Theory

During the 1920's and 30's the Serbian mathematician Milutin Milankovitch studied cyclical variations in three elements of the Earth–Sun geometry: eccentricity, precession, and obliquity, and was able to calculate the Earth's solar radiation history for the past 650 ka (Milankovitch, 1969). Berger (1978, 1980) accurately determined the periodicities of the three orbital variations.

Eccentricity—The Earth's orbit around the Sun is an ellipse; this results in the seasons. The eccentricity of the Earth's orbit periodically departs further from a circle and then reverts to almost true circularity. Periodicities are located around 413, 95, 123, and 100 ka. Secondary peaks appear to be located around 50 and 53 ka. There are further important periodicities at 1.23, 2.04, and 3.4 ma (Schwarzacher, 1991).

Precession—Precession refers to variation in time of year at which the Earth is nearest the Sun (perihelion). This variation is caused by the Earth wobbling like a top and swiveling on its axis. Periodicities of 23,000, 22,400, 18,980, and 19,610 yr are recognized and often simplified to two periods of 19 and 23 ka. Secondary peaks are also located around 30 and 15 ka.

Obliquity—The tilt of the Earth's axis of rotation (varying between $21°39'$ to $24°36'$), which affects seasonal contrasts, is referred to as obliquity. The greater the tilt, the more pronounced is the difference between winter and summer. The primary periodicity is 41 ka with a secondary periodicity of approximately 56 ka.

Although orbital variations do not change the solar constant appreciably (*i.e.*, amount of solar energy normally received on a unit area at the top of the Earth's atmosphere), they significantly alter the latitudinal and seasonal distribution of solar radiation received by the Earth. This variation in insolation has a direct influence on the Earth's climate (Imbrie *et al.*, 1984).

Milankovitch Band Cyclicity

Natural systems such as those operating within active sedimentary basins are often too complex to be realistically modeled. However, systems may be strongly influenced or "forced" by an external control parameter to the extent that their evolution may become predictable if the external influence is understood. The geological record demonstrates that some sedimentary systems have been forced by variations in the Earth's climate which are in turn linked to the mechanics of the Earth's orbit. The evidence for this comes in the form of cyclic sedimentary phenomena within the "Milankovitch band" of astronomical frequencies.

Over the past decade there has been a substantial increase in the literature on Milankovitch band sedimentary phenomena (Algeo and Wilkinson, 1988). Milankovitch band frequencies have been recorded from cyclical alluvial deposits (Olsen, 1990), lake deposits (Bradley, 1929), evaporites (Williams, 1991), shallow-marine clastics (Clifton, 1981), shallow marine carbonates (Fischer, 1964), and pelagic sediments (de Boer, 1982).

Detection and Analysis of Cyclicity

To assess patterns and cyclicity in the stratigraphic record in a statistically meaningful manner, data are needed which can be treated as a time series.

Strict equivalence with orthodox time series is often complicated by variations in sediment accumulation rates and discontinuities caused by erosional breaks and periods of non-sedimentation. Time series may be subjected to power spectral analysis to detect regular cyclicity hidden in "noise" (Davis, 1973; Weedon, 1989).

Geologists have used power spectral analysis to identify sunspot cycles in varved sediments (Williams, 1981; Williams and Sonett, 1985) and neap/spring cycles in tidal sediments (Williams, 1989), as well as astronomically forced Milankovitch cycles (Fischer, 1964). A wide variety of data has been subjected to spectral analysis to detect cyclicity: geochemical (*e.g.,* Cubitt, 1979; Williams, 1991), isotopic (Hays *et al.,* 1976), natural radioactivity (Kelly, 1992), and sedimentary facies (Olsen, 1990).

Relative frequency is a critical factor in determining if a sedimentary cycle is related to Milankovitch band phenomena. If power spectral analysis reveals a number of cycles, frequency ratios may be used to recognize the Milankovitch spectrum even without an absolute time scale. This methodology is complicated because the amplitude and frequency of orbital parameters may vary with both latitude and geological time. If sedimentary cyclicity and Milankovitch band frequencies can be correlated, a time frame can be established and accurate estimates of sediment accumulation rates made (House, 1985; Park and Herbert, 1987; Schwarzacher, 1987, 1991; Algeo and Wilkinson, 1988).

Precession–eccentricity-dominated patterns are most pronounced in middle to low latitudes, whereas the climatic effect of obliquity oscillation is most pronounced in (but not confined to) polar regions (Ruddiman and MacIntyre, 1981; Turcotte and Bernthal, 1984; Fischer *et al.,* 1990). Precambrian glaciogenic sequences contain evidence of marked seasonal unequability of climate (Williams, 1975, 1981), possibly caused by increased obliquity of the ecliptic in the late Precambrian which would weaken climatic zonation and allow ice sheets to form in low latitudes.

Calculating orbital parameters (both frequency and amplitude) backwards through time over hundreds of millions of years is not straightforward, as there has apparently been some variation in the orbital behavior of the Earth through geological time (Lambeck, 1978; Walker and Zahnle, 1986; Berger *et al.,* 1989*a,b*). Both the obliquity and precession cycles are influenced by the Earth–Moon distance which has increased with time (Walker and Zahnle, 1986), leading the frequency of both elements to increase with time (Berger *et al.,* 1989*a,b*; Collier *et al.,* 1990; Williams, 1991). The amplitude of the obliquity cycle has also increased (Collier *et al.,* 1990).

Ice Ages

Milankovitch proposed that Pleistocene glacial climatic episodes were initiated when the Earth–Sun geometry favored minimum summer insolation at high northern latitudes. Sediment and ice cores have shown that for much of the Earth's recent past, climate has fluctuated periodically and the astronomical or Milankovitch theory is a valid explanation of the longer scale of environmental change (Hays et al., 1976; Imbrie et al., 1984). Furthermore, astronomical forcing of climate may not be restricted to planet Earth (Cutts and Lewis, 1982).

Ice ages are probably initiated by interactions between land configuration and altitude, ocean characteristics, polar position, and oceanic circulation (Steiner and Grillmair, 1973). Major continental glaciations probably require that continental masses be positioned over the poles or that continents encircle polar seas (Crowell and Frakes, 1970; Frakes, 1979).

The world has probably experienced glaciation since the Precambrian, with oscillations between major continental ice ages and intervening periods when glaciations were restricted to upland areas (Smith et al., 1973; Harland and Herod, 1975; Sugden and John, 1979; Hardie et al., 1986; Horbury, 1989). There is wide consensus that global climatic conditions during the Paleozoic were not consistent (Frakes, 1979; John, 1979; Spjeldnaes, 1981). The record of glaciation through the Paleozoic can be correlated with movement of the Gondwanaland continents across the south pole (Fairbridge, 1973; Crowell, 1978; Frakes, 1979; Caputo and Crowell, 1985; Veevers and Powell, 1987).

Below-average global temperatures prevailed during the Ordovician and early Silurian and from the Devonian–Carboniferous boundary to the early Permian when there was a sudden increase in global temperature (Frakes, 1979). During globally cool intervals, climatic zones shifted toward the equator, latitudinal gradients were steep, and conditions became favorable for ice sheet growth. When global temperatures were above average, such as during the late Silurian to Devonian, climatic zones shifted polewards, but continents clustered at the pole would have experienced cool climatic conditions. The Permo-Carboniferous ice sheet grew because the pole was located over the highlands of Antarctica and subduction of the oceanic paleo-Pacific plate beneath the Gondwana plate created an alpine mountain range (Frakes et al., 1971).

Like their Pleistocene counterparts, the older Phanerozoic ice ages also experienced glacial-interglacial fluctuations (Bond and Stocklmayer, 1967; Spencer, 1971), and it is possible that such fluctuations had a common cause.

Eustasy

Change in the volume of continental ice sheets is often cited as a cause of global change in sea level (Vail et al., 1977; Veevers and Powell, 1987). In its simplest form the glacio-eustatic theory suggests that sea level oscillates in response to the quantity of water stored in ice caps during glaciations and interglacials; transgressions during interglacials are succeeded by regressions during the glacials. Such changes are limited to but not inherent in "ice-house" periods which are characterized by the development of polar ice caps. The rates of change are potentially high -0.01 to 0.1 m.yr^{-1} (Pitman, 1978), although the actual magnitude of sea-level oscillations was less dramatic in "greenhouse" times such as the Jurassic and Cretaceous than in the Pleistocene (c. 100 m). The effect of relatively rapid sea-level rises on low-gradient (typically $< 1 : 1000$) continental shelves can be dramatic with transgression rates on the order of 10 m.yr^{-1} (Evans, 1979).

Eustatic cycles are strongly asymmetric, gradually falling to lowstand conditions followed by an abrupt rise to highstand (Vail *et al.*, 1977). Isotope data from equatorial cores can be matched with detailed sea-level curves (Chappell and Shackleton, 1986), demonstrating that a glacio-eustatic sea-level curve corresponds to the 100-ka cycle of rapid ablation followed by gradual ice sheet growth. A comparison of the sea-level curves of Vail *et al.* (1977) with the paleoclimatic data for the Cenozoic based on isotope data suggests a coincidence of highstands with periods of global warming (Haq *et al.*, 1977; Miller *et al.*, 1987).

Sequence Stratigraphy

Using the terminology of sequence stratigraphy, some fifth- and fourth-order sequences (parasequences or paracycles) can be interpreted as the result of astronomically forced climatic variation and related eustasy. Turcotte and Willeman (1983) and Collier *et al.* (1990) demonstrated that sinusoidal sea-level fluctuations superimposed on a constant rate of tectonic subsidence simulates cyclic stratigraphy. Milankovitch periodicities may be used to estimate the magnitude and frequency of glacio-eustatic sea-level changes (Grotzinger, 1986; Read *et al.*, 1986; Koerschner and Read, 1989; Collier *et al.*, 1990).

Paracycles or parasequences are frequently described from shallow marine carbonate systems (Wong and Oldershaw, 1980; Dorobek, 1986; Grotzinger, 1986; Goldhammer *et al.*, 1987; Bronner, 1988; Goldhammer and Harris, 1989; Horbury, 1989; Koerschner and Read, 1989; Playford *et al.*, 1989; Read, 1989). In a study of Cretaceous limestones, Gilbert (1895) suggested that carbonate cyclicity may be induced astronomically and the most likely cause of regular bedding was precession of the equinoxes. The cycles are often packaged into thicker cycles consisting of five individual cycles (five precession cycles are contained within one eccentricity cycle).

Carbonate paracycles often are regressive shallowing-upward sequences dominated by subtidal/tidal-flat sediments. Individual cycles are 1–10 m thick, determined by the amplitude of sea-level fluctuations and duration of high stands (Read *et al.*, 1986). Many ancient carbonate shelves were aggraded mature features (Koerschner and Read, 1989) and were susceptible to floods by low-amplitude sea-level oscillations. Sedimentation rates were able to either keep up with relative sea-level rise or the sediment/water interface remained above wavebase (Read *et al.*, 1986).

Conformable contacts between paracycles indicate that sea-level fall rates roughly equaled or were less than subsidence. Erosional contacts between cycles indicate that sea-level fall rates exceeded subsidence rates. Disconformity-bounded cycles reflect high-amplitude sea-level oscillations, which inhibited progradation of tidal flats and allowed development of paleosol profiles (Hardie *et al.*, 1986; Goldstein, 1988; Horbury, 1989).

Parasequences or "cyclothems" are well documented from Upper Paleozoic sediments of the North American Midcontinent (Weller, 1930; Moore, 1931; Wanless and Shepard, 1936; Moore, 1959; Moore and Merriam, 1959; Merriam, 1963, 1964; Cubitt, 1979; Heckel, 1977, 1986, 1990). The cyclothems are generally 5 to 25 m thick and consist of a thin transgressive limestone, thin offshore dark phosphatic shale, and thick regressive limestone. Most cyclothems are separated by thin clastic terrestrial deposits with paleosols. An "ideal" cycle is thought to record a relatively rapid transgression followed by a more gradual regression (Heckel, 1990), possibly reflecting the asymmetric character of ice cap ablation-growth fluctuations. The lateral extent and estimated frequencies of the cyclothems suggest that they are induced by eustasy related to waxing and waning of the Gondwana ice cap (Crowell, 1978; Veevers and Powell, 1987).

Parasequences within clastic systems are less frequently documented in the literature. Pleistocene clastic sequences are interpreted by Clifton (1981) and Clifton *et al.* (1988) as caused by eustatic sea-level changes in a relatively rapidly and uniformly subsiding basin. The number of cycles approximately equals the number of major sea-level cycles in the Pleistocene indicated by oxygen isotope data from foraminifera in deep sea cores. A cycle starts with a scoured surface and shoals upward, ending with nonmarine deposition.

Plint (1991) described Cretaceous shallow-marine clastic parasequences thought to have been generated by high-frequency relative sea-level oscillations. Each parasequence consists of shale or bioturbated siltstone passing upward into cross-laminated and cross-bedded sandstones and locally capped by a pebble bed. The parasequences are 5–10 m thick and can be traced regionally. The upward-coarsening units can be interpreted as a shelf-aggradation response to shoreface progradation when space was created by a relative rise in sea level. During the succeeding lowstand, the effective wave base was lowered sufficiently for erosion and high-energy transportation of gravelly material. The presence of pebble beds suggests short-term relative fall. The parasequences have an apparent periodicity of 100 ka which suggests glacio-eustasy related to orbital forcing (Plint, 1991). Similar clastic parasequences have been observed by one of the authors (SBK) in the Jurassic of the North Sea and the Cretaceous of the Middle East.

Pelagic Cycles

There is evidence of orbitally controlled cycles in pelagic sediments in the form of alternating limestone-marl rhythms (de Boer, 1982; Schwarzacher and Fischer, 1982; Arthur *et al.*, 1984; Hattin, 1986; Herbert *et al.*, 1986; Herbert and Fischer, 1986; Weedon, 1986, 1989; Pratt, 1988). The cyclicity probably resulted from periodic fluctuations of the lysocline and the carbonate compensation depth (CCD) which changed the area of the seafloor affected by carbonate dissolution and altered the carbonate/clay ratio. Cyclic black shale–chert couplets were deposited as siliceous ooze during periods of upwelling and high organic productivity and black mudstones during times of reduced oxygenation of bottom waters (Roberts and Mitterer, 1988).

Cyclicity in pelagic environments may also be related to climatic change and eustasy. Highstands tend to result in shoaling of the CCD and increased dissolution in the deep basins (Haq, 1991). Elevated sea levels also cause climatic equability, weakening of latitudinal and vertical thermal gradients, and reduction of bottom-water activity. During sea-level falls, terrigenous as well as carbonate sediments bypass the shelves into the basins, resulting in increased carbonate in the oceans and depressed CCD. Lowstands are likely to be characterized by sharper temperature gradients due to climatic inequability and increased current activity (Haq, 1991).

Nonmarine Environments

Variations in insolation can induce climatic change (and potentially, sedimentary cyclicity) in several manners apart from the glacio-eustatic control. Insolation controls land–sea thermal contrasts that promote air mass uplift and condensation over heated continents and subsidence and evaporation over cooler oceans. This affects rates of evaporation/precipitation, although the feedback effects of albedo can also be influential (Barron *et al.*, 1980), which in turn influences sedimentation patterns and styles in the nonmarine realm. Variations in runoff may potentially be highly significant as regards the resultant facies and architecture of both alluvial and lacustrine deposits.

With fluctuations in the Earth's insolation signal, concomitant fluctuations might be expected in alluvial discharge. During cold/dry periods, rivers carry reduced discharges which are reflected in relatively small sediment loads. During warm/wet periods, in contrast, rivers are able to carry greater loads as precipitation and runoff are increased. By studying the size of fluvial channel bodies and sedimentary structures contained within them and making basic assumptions concerning the paleohydrology of alluvial channels, Olsen (1990) and Kelly (1992) have demonstrated that orbitally forced variations of fluvial discharge occurred during the Devonian. Astronomically forced variations in discharge influence parameters such as channel size, density, and frequency, as well as the character of flood-plain deposits (Kelly, 1992).

Sequences of lacustrine sediments can be strongly cyclical with frequencies within the Milankovitch band, with precession and eccentricity cycles dominating (Bradley, 1929; Olsen, 1986, 1990; Astin, 1990; Kelly, 1992). Lake deposits often exhibit transgressive–regressive cycles in much the same manner as marine deposits (Donovan, 1980; Olsen, 1986, 1990; Astin, 1990), as well as both glacial and non-glacial varves. In instances cyclical lacustrine sediments can be calibrated by the presence of varves which are assumed to be annual in origin (Olsen, 1990).

Summary

1. During the Earth's recent past, climate has apparently fluctuated cyclically with periodicities of about 20, 40, and 100 ka, as predicted by Milutin Milankovitch (Imbrie *et al.*, 1984). These periodicities represent periodic changes in solar radiation received by the Earth because of changes in the distance between the Earth and the Sun caused by variation in the Earth's orbit. Periodicities in precession, obliquity, and eccentricity account for the three main Milankovitch cycles (Imbrie *et al.*, 1984).

2. Examination of ancient sediments of all ages back into the Precambrian suggests that astronomical cycles have been a constant feature of planet Earth and are not restricted to "icehouse" periods of the geological past, but also occur during "greenhouse" times when there is little evidence for large polar ice caps. Milankovitch band climatic fluctuations have been recorded in a wide range of sedimentary environments ranging from alluvial to pelagic settings.

3. Astronomically forced glacial eustasy is a global process that requires the presence of large ice caps that wax and wane in the higher latitudes, and can result in distinct stratigraphic units of extremely widespread extent and correlatability. Milankovitch band parasequences occur in both carbonate and clastic shelf systems (Grotzinger, 1986; Plint, 1991). Parasequences are frequently asymmetric in character. This is possibly a reflection of the asymmetric nature of sea-level fluctuations which are characterized by gradual regressions and rapid transgressions. The cyclothemic succession described from the Upper Paleozoic of the Midcontinent of North America represents an excellent example of a glacio-eustatically driven sedimentary system (Heckel, 1990).

4. Milankovitch cycles provide an independent method of calibrating sediment accumulation rates (House, 1985; Park and Herbert, 1987; Schwarzacher, 1991). Because the average frequencies of Milankovitch cycles tend to increase into areas with high subsidence rates, this

also allows the effects of differential basinal subsidence to be studied (Koerschner and Read, 1989; Olsen, 1990).

References

Algeo, T.J. and Wilkinson, B.H., 1988, Periodicity of mesoscale sedimentary cycles and the role of Milankovitch orbital modulation: *Jour. Geol.*, v. 96, p. 313–322.

Arthur, M.A., Dean, W.E., Bottjer, D. and Scholle, P.A., 1984, Rhythmic bedding in Mesozoic–Cenozoic pelagic carbonate sequences: The primary and diagenetic origin of Milankovitch-like cycles, *in* Berger, A.L., Imbrie, J., Kukla, G. and Saltzman, B., eds., *Milankovitch and Climate:* Reidel, Amsterdam, p. 191–222.

Astin, T.R., 1990, The Devonian lacustrine sediments of Orkney, Scotland; Implications for climate cyclicity, basin structure and maturation history: *Jour. Geol. Soc.*, v. 147, p. 141–151.

Barron, E.J., Sloan, J.L. and Harrison, C.G.A., 1980, Potential significance of land–sea distribution and surface albedo variations as a climatic factor: 180 my to present: *Palaeogeography, Palaeoclimatology, Palaeoecology*, v. 30, p. 17–40.

Berger, A.L., 1978, Long-term variations of daily insolation and Quaternary climatic changes: *Jour. Atmospheric Sci.*, v. 5, p. 2362–2367.

Berger, A.L., 1980, The Milankovitch astronomical theory of paleoclimates: A modern review, *in* Beer, A., Pounds, K. and Beer, P., eds., *Vistas in Astronomy:* Pergamon Press, Oxford, p. 103–122.

Berger, A.L., Loutre, M.F. and Dehant, V., 1989*a*, Influence of the changing lunar orbit on the astronomical frequencies of pre-Quaternary insolation patterns: *Paleoceanography*, v. 4, p. 555–564.

Berger, A., Loutre, M.F. and Dehant, V., 1989*b*, Pre-Quaternary Milankovitch frequencies: *Nature*, v. 342, p. 133.

de Boer, P.L., 1982, Cyclicity and the storage of organic matter in middle Cretaceous pelagic sediments, *in* Einsele, G. and Seilacher, A., eds., *Cyclic and Event Stratification:* Springer-Verlag, Heidelberg, p. 456–475.

Bond, G. and Stocklmayer, V.R.C., 1967, Possible ice margin fluctuations in the Dwyka series in Rhodesia: *Palaeogeography, Palaeoclimatology, Palaeoecology*, v. 3, p. 433–446.

Bradley, W.H., 1929, *The varves and climate of the Green River Epoch:* U.S. Geological Survey, Prof. Paper 645, p. 1–108.

Bronner, R.L., 1988, Cyclicity in Upper Mississippian Bangor Limestone, Blount County, Alabama: *Bulletin Am. Assoc. Petroleum Geologists*, v. 72, p. 166.

Caputo, M.V. and Crowell, J.C, 1985, Migration of glacial centers across Gondwana during the Paleozoic era: *Geol. Soc. Amer. Bull.*, v. 96, p. 1020–1036.

Chappell, J. and Shackleton, N.J., 1986, Oxygen isotopes and sea level: *Nature*, v. 324, p. 137–140.

Clifton, H.E., 1981, Progradational sequences in Miocene shoreline deposits, southeastern Caliente Range, California: *Jour. Sed. Pet.*, v. 51, p. 165–184.

Clifton, H.E., Hunter, R.E. and Gardner, J.V., 1988, Analysis of eustatic, tectonic, and sedimentologic influences on transgressive and regressive cycles in the Upper Cenozoic Merced Formation, San Francisco, California, *in* Kleispehn, K.L. and Paola, C., eds., *New Perspectives in Basin Analysis:* Springer–Verlag, New York, p. 109–128.

Collier, R.E.L, Leeder, M.R. and Maynard, J.R., 1990, Transgressions and regressions: A model for the influence of tectonic subsidence, deposition and eustasy, with application to Quaternary and Carboniferous examples: *Geol. Mag.*, v. 127, p. 117–128.

Crowell, J.C., 1978, Gondwanan glaciation, cyclothems, continental positioning, and climatic change: *Am. Jour. Sci.*, v. 278, p. 1345–1372.

Crowell, J.C. and Frakes, L.A., 1970, Phanerozoic ice ages and the cause of ice ages: *Am. Jour. Sci.*, v. 268, p. 193–224.

Cubitt, J.M., 1979, Geochemistry, mineralogy and petrology of Upper Paleozoic shales of Kansas: *Kansas Geol. Survey Bull.*, no. 217, 117 pp.

Cutts, J.A. and Lewis, B.H., 1982, Models of climate cycles recorded in Martian polar layered deposits: *Icarus*, v. 50, p. 216–244.

Davis, J.C., 1973, *Statistics and Data Analysis in Geology:* John Wiley & Sons, New York, 550 pp.

Donovan, R.N., 1980, Lacustrine cycles, fish ecology and stratigraphic zonation in the Middle Devonian of Caithness: *Scott. Jour. Geol.*, v. 16, p. 35–60.

Dorobek, S.L., 1986, Cyclic platform dolomites of Devonian Jefferson Formation, Montana and Idaho: *Bulletin Am. Assoc. Petroleum Geologists*, v. 70, p. 583.

Evans, G., 1979, Quaternary transgressions and regressions: *Jour. Geol. Soc.*, v. 136, p. 125–132.

Fairbridge, R.W., 1973, Glaciation and plate migration, *in* Tarling, D.H. and Runcorn, S.K., eds., *Implications of Continental Drift to the Earth Sciences:* Academic Press, London, p. 503–515.

Fischer, A.G., 1964, The Lofer cyclothems of the Alpine Triassic, *in* Merriam, D.F., ed., *Symposium on Cyclic Sedimentation:* Kansas Geol. Survey Bull., no. 169, p. 107–149.

Fischer, A.G., de Boer, P.L. and Premoli-Silva, I., 1990, Cyclostratigraphy, *in* Ginsburg, R.N. and Beaudoin, B., eds., *Cretaceous Resources, Events and Rhythms:* Kluwer, Dordrecht, p. 139–172.

Frakes, L.A., 1979, *Climates Throughout Geological Time:* Elsevier, Amsterdam, 310 pp.

Frakes, L.A., Matthews, J.L. and Crowell, J.C., 1971, Late Paleozoic glaciation: Part III, Antarctica: *Geol. Soc. Amer. Bull.*, v. 82, p. 1581–1604.

Gilbert, G.K., 1895, Sedimentary measurement of geological time: *Jour. Geol.*, v. 3, p. 121–127.

Goldhammer, R.K., Dunn, P.A. and Hardie, L.A., 1987, High-frequency glacio-eustatic sea-level oscillations with Milankovitch characteristics recorded in Middle Triassic platform carbonates in northern Italy: *Am. Jour. Sci.*, v. 287, p. 853–892.

Goldhammer, R.K. and Harris, M.T., 1989, Eustatic controls on the stratigraphy and geometry of the Latemar Buildup (Middle Triassic), the Dolomites of Northern Italy, *in* SEPM Spec. Publ. 44: *Controls on Carbonate Platform and Basin Development*, p. 325–338.

Goldstein, H., 1988, Paleosols of Late Pennsylvanian cyclic strata, New Mexico: *Sedimentology*, v. 35, p. 777–803.

Grotzinger, J.P., 1986, Upward shallowing platform cycles: A response to 2.2 billion years of (Milankovitch band) sea-level oscillations: *Paleoceanography*, v. 1, p. 403–416.

Haq, B.U., 1991, Sequence stratigraphy, sea-level change and significance for the deep sea, *in* MacDonald, D.I.M., ed., *Sedimentation, Tectonics and Eustasy:* Int. Ass. Sed., Spec. Publ. 12, p. 3–39.

Haq, B.U., Premoli-Silva, I. and Lohmann, G.P., 1977, Calcareous plankton paleobiogeographic evidence for major climatic fluctuations in the Early Cenozoic Atlantic Ocean: *Jour. Geophysical Research*, v. 82, p. 3861–3876.

Hardie, L.A., Bosellini, A. and Goldhammer, R.K., 1986, Repeated subaerial exposure of subtidal carbonate platforms. Triassic, Northern Italy: Evidence for high frequency sea-level oscillations on a 10^4 year scale: *Paleoceanography*, v. 1, p. 447–459.

Harland, W.B. and Herod, K.N., 1975, Glaciations through time, *in* Wright, A.E. and Moseley, F., eds., *Ice Ages: Ancient and Modern:* Seel House Press, Liverpool, p. 189–216.

Hattin, D.E., 1986, Interregional model for the deposition of Upper Cretaceous pelagic rhythmites, US Western Interior: *Paleoceanography*, v. 1, p. 483–494.

Hays, J.D., Imbrie, J. and Shackleton, N.J., 1976, Variations in the Earth's orbit: Pacemaker of the Ice Ages: *Science*, v. 194, p. 1121–1132.

Heckel, P.H., 1977, Origin of phosphatic black shale facies in Pennsylvanian cyclothems of Midcontinent North America: *Bulletin Am. Assoc. Petroleum Geologists*, v. 61, p. 1045–1068.

Heckel, P.H., 1986, Sea-level curve for Pennsylvanian eustatic marine transgressive–regressive depositional cycles along Mid-continent outcrop belt, North America: *Geology*, v. 14, p. 330–334.

Heckel, P.H., 1990, Evidence for global (glacial-eustatic) control over Upper Carboniferous (Pennsylvanian) cyclothems in Midcontinent North America, *in* Hardman, R.F.P. and Brooks, J., eds., *Tectonic Events Responsible for Britain's Oil and Gas Reserves:* Geol. Soc., Spec. Publ. 55, p. 35–47.

Herbert, T.D. and Fischer, A.G., 1986, Milankovitch climatic origin of mid-Cretaceous black shale rhythms in central Italy: *Nature*, v. 321, p. 739–743.

Herbert, T.D., Stallard, A.F. and Fischer, A.G., 1986, Anoxic events, productivity rhythms and the orbital signature in a mid-Cretaceous deep sea sequence from central Italy: *Paleoceanography*, v. 1, p. 495–507.

Horbury, A.D., 1989, The relative role of tectonism and eustasy in the deposition of the Urswick Limestone in South Cumbria and North Lancashire, *in* Arthurton, R.S., Gutteridge, P. and

Nolan, S.C., eds., *The Role of Tectonics in Devonian and Carboniferous Sedimentation in the British Isles:* Yorks. Geol. Soc., Occ. Publ. 6, p. 153–169.

House, M.R., 1985, A new approach to an absolute timescale from measurements of orbital cycles and sedimentary microrhythms: *Nature*, v. 315, p. 721–725.

Imbrie, J., Hays, J.D., Martinson, D.G., McIntyre, A., Mix, A.C., Morley, J.J., Pisias, N.G., Prell, W. and Shackleton, N.J., 1984, The orbital theory of Pleistocene climate: Support from a revised chronology of the O–18 record, *in* Berger, A.L., Imbrie, J., Hays, J., Kukla, G. and Saltzman, B., eds., *Milankovitch and Climate:* D. Riedel, Hingham, Mass., p. 269–305.

John, B.S., 1979, *The Winters of the World:* David & Charles, London, 256 pp.

Kelly, S.B., 1992, Milankovitch cyclicity recorded from Devonian non-marine sediments: *Terra Nova*, v. 4, p. 578–584.

Koerschner, W.F., III and Read, J.F., 1989, Field and modelling studies of Cambrian carbonate cycles, Virginia Appalachians: *Jour. Sed. Pet.*, v. 59, p. 654–687.

Lambeck, K., 1978, The Earth's palaeorotation, *in* Brosche, P. and Sunderman, J., eds., *Tidal Friction and the Earth's Rotation:* Springer–Verlag, Heidelberg, p. 145–153.

Merriam, D.F., 1963, The geological history of Kansas: *Kansas Geol. Survey Bull.*, no. 162, 317 pp.

Merriam, D.F., 1964, Symposium on cyclic sedimentation: *Kansas Geol. Survey Bull.*, no. 169, 636 pp.

Milankovitch, M., 1969, *Canon of Insolation and the Ice Age Theory:* Israel Progr. Sci. Transl., Jerusalem, 484 pp.

Miller, K.G., Fairbanks, R.G. and Mountain, G.S., 1987, Tertiary oxygen isotope synthesis, sealevel history and continental margin erosion: *Paleoceanography*, v. 2, p. 1–19.

Moore, D., 1959, Role of deltas in the formation of some British Lower Carboniferous cyclothems: *Jour. Geol.*, v. 67, p. 522–539.

Moore, R.C., 1931, Pennsylvanian cycles in the northern Mid-Continent region: *Illinois Geol. Survey Bull.*, v. 60, p. 247–257.

Moore, R.C. and Merriam, D.F., 1959, *Kansas Field Conference—1959:* Kansas Geol. Survey, Lawrence, Kan., p. 1–55 [Guidebook].

Olsen, H., 1990, Astronomical forcing of meandering river behaviour: Milankovitch cycles in Devonian of East Greenland: *Palaeogeography, Palaeoclimatology, Palaeoecology*, v. 79, p. 99–115.

Olsen, P.E., 1986, A 40 million year lake record of early Mesozoic orbital climatic forcing: *Science*, v. 234, p. 842–848.

Olsen, P.E., 1990, Tectonic, climatic, and biotic modulations of lacustrine ecosystems—Examples from Newark Supergroup of Eastern North America, *in* Katz, B.J., ed., *Lacustrine Basin Exploration: Case Studies and Modern Analogs:* Am. Assoc. Petroleum Geologists, Memoir 50, p. 209–224.

Park, J. and Herbert, T.D., 1987, Hunting for paleoclimatic periodicities in a geologic time series with an uncertain time scale: *Jour. Geophysical Research*, v. 92, p. 14027–14040.

Pitman, W.C., III, 1978, Relationship between eustasy and stratigraphic sequences of passive margins: *Geol. Soc. Amer. Bull.*, v. 89, p. 1389–1403.

Playford, P.E., Hurley, N.F., Kerans, C. and Middleton, M.F., 1989, Reefal platform development, Devonian of the Canning Basin, Western Australia*in* SEPM Spec. Publ. 44: *Controls on Carbonate Platform and Basin Development*, p. 193–200.

Plint, A.G., 1991, High-frequency relative sea-level oscillations in Upper Cretaceous shelf clastics of the Alberta foreland basin: Possible evidence for a glacio-eustatic control?, *in* MacDonald, D.I.M., ed., *Sedimentation, Tectonics and Eustasy:* Int. Ass. Sed., Spec. Publ. 12, p. 409–428.

Pratt, L.M., 1988, Importance of geochemical/isotopic events as stratigraphic markers in cyclic Cretaceous strata: *Bulletin Am. Assoc. Petroleum Geologists*, v. 72, p. 237.

Read, J.F., 1989, Controls on evolution of Cambrian–Ordovician passive margin, U.S. Appalachians, *in* SEPM Spec. Publ. 44: *Controls on Carbonate Platform and Basin Development*, p. 157–165.

Read, J.F., Grotzinger, J.P., Bova, J.A. and Koerschner, W.F., 1986, Models for generation of carbonate cycles: *Geology*, v. 14, p. 107–110.

Roberts, C.T. and Mitterer, R.M., 1988, Laminated black shale–Chert cyclicity in Woodford Formation (Upper Devonian of southern Mid-Continent): *Bulletin Am. Assoc. Petroleum Geologists*, v. 72, p. 240–241.

Ruddiman, W.F. and McIntyre, A., 1981, Oceanic mechanisms for the amplification of the 23,000 year ice-volume cycle: *Science*, v. 229, p. 617–626.

Schwarzacher, W., 1987, Astronomical cycles for measuring geological time: *Modern Geology*, v. 11, p. 375–381.

Schwarzacher, W., 1991, Timing and correlation. Milankovitch cycles and the measurement of time, *in* Einsele, G., Ricken, W. and Seilacher, A., eds., *Cycles and Events in Stratigraphy*: Springer–Verlag, Heidelberg, p. 855–863.

Schwarzacher, W. and Fischer, A.G., 1982, Limestone–shale bedding and perturbations of the Earth's orbit, *in* Einsele, G. and Seilacher, A., eds., *Cyclic and Event Stratification*: Springer–Verlag, Heidelberg, p. 72–95.

Smith, A.G., Briden, J.C. and Drewry, G.E., 1973, Phanerozoic world maps: *Spec. Pap. Palaeont.*, v. 12, p. 1–42.

Spencer, A.M., 1971, *Late Precambrian Glaciations in Scotland*: Geol. Soc., Mem. 6, 98 pp.

Spjeldnaes, N., 1981, Lower Palaeozoic palaeoclimatology, *in* Holland, G.H., ed., *Lower Palaeozoic of the Middle East Eastern and Southern Africa, and Antarctica*: John Wiley & Sons, New York, p. 199–256.

Steiner, J. and Grillmair, E., 1973, Possible galactic causes for periodic and episodic glaciations: *Geol. Soc. Amer. Bull.*, v. 84, p. 1002–1018.

Sugden, D.E. and John, B.S., 1979, *Glaciers and Landscape*: Arnold, New York, 376 pp.

Turcotte, D.L. and Bernthal, M.J., 1984, Synthetic coral-reef terraces and variations of Quaternary sea level: *Earth and Planetary Sci. Letters*, v. 70, p. 121–128.

Turcotte, D.L. and Willeman, R.J., 1983, Synthetic cyclic stratigraphy: *Earth and Planetary Sci. Letters*, v. 63, p. 89–96.

Vail, P.R., Mitchum, R.M., Jr., Todd, R.G., Widmier, J.M., Thompson, S., Sangree, J.B., Dubb, J.N. and Haslid, W.G., 1977, Seismic stratigraphy and global changes of sea level, Parts 1–6, *in* Payton, C.E., ed., *Seismic Stratigraphy—Applications to Hydrocarbon Research*: Am. Assoc. Petroleum Geologists, Memoir 26, p. 49–133.

Veevers, J.J. and Powell, C.Mc.A., 1987, Late Paleozoic glacial episodes in Gondwanaland reflected in transgressive–regressive sequences in EuroAmerica: *Geol. Soc. Amer. Bull.*, v. 98, p. 475–487.

Walker, J.C.G. and Zahnle, K.J., 1986, Lunar nodal tide and distance to the Moon during the Precambrian: *Nature*, v. 320, p. 600–602.

Wanless, H.R. and Shepard, F.P., 1936, Sea level and climatic changes related to the late Paleozoic cycles: *Geol. Soc. Amer. Bull.*, v. 47, p. 1177–1206.

Weedon, G.P., 1986, Hemipelagic shelf sedimentation and climatic cycles: The basal Jurassic (Blue Lias) of S. Britain: *Earth and Planetary Sci. Letters*, v. 76, p. 321–335.

Weedon, G.P., 1989, The detection and illustration of regular sedimentary cycles using Walsh power spectra and filtering, with examples from the Lias of Switzerland: *Jour. Geol. Soc.*, v. 146, p. 133–144.

Weller, J.M., 1930, Cyclical sedimentation of the Pennsylvanian period and its significance: *Jour. Geol.*, v. 38, p. 97–135.

Williams, G.E., 1975, Late Precambrian glacial climate and the Earth's obliquity: *Geol. Mag.*, v. 112, p. 441–465.

Williams, G.E., 1981, Sunspot periods in the late Precambrian glacial climate and solar–planetary relations: *Nature*, v. 291, p. 624–628.

Williams, G.E., 1989, Late Precambrian tidal rhythmites in South Australia and the history of the Earth's rotation: *Jour. Geol. Soc.*, v. 146, p. 97–111.

Williams, G.E., 1991, Milankovitch-band cyclicity in bedded halite deposits contemporaneous with Late Ordovician–Early Silurian glaciation, Canning Basin, Western Australia: *Earth and Planetary Sci. Letters*, v. 103, p. 143–155.

Williams, G.E. and Sonett, C.P., 1985, Solar signature in sedimentary cycles from the late Precambrian Elatina Formation, Australia: *Nature*, v. 318, p. 523–527.

Wong, P.K. and Oldershaw, A.E., 1980, Causes of cyclicity in reef interior sediments, Kaybob Reef, Alberta: *Bull. Can. Pet. Geol.*, v. 28, p. 411–424.

12

CAN THE GINSBURG MODEL GENERATE CYCLES?

W. Schwarzacher

The Ginsburg model of carbonate accumulation is an often-quoted mechanism for generating so-called autocycles. It is shown that the model does not represent a self-oscillating system; oscillations can only be generated if at least two critical parameters controlling sedimentation are introduced.

The Ginsburg Model

The Ginsburg model is a conceptual model which tries to explain the behavior of some carbonate shelves that undergo continuous tectonic subsidence and that carry on their surface a very active carbonate factory. Unfortunately the original model has only been published in abstract form—the relevant part of which is quoted here in full (Ginsburg, 1971, p. 340):

> "The Florida Bay lagoon and the tidal flats of the Bahamas and Persian Gulf are traps for fine sediment produced on the large adjacent open platforms or shelves. The extensive source areas produce carbonate mud by precipitation and by the disintegration of organic skeletons. The carbonate mud moves shoreward by wind-driven, tidal or estuarinelike circulation, and deposition is accelerated and stabilized by marine plants and animals.
>
> Because the open marine source areas are many times larger than the nearshore traps, seaward progradation of the wedge of sediments is inevitable. This seaward progradation gives a regressive cycle from open marine shelf or platform to supratidal flat. As the shoreline progrades seaward the size of the open marine source area decreases; eventually reduced production of

mud no longer exceeds slow continuous subsidence and a new transgression begins. When the source area expands so that production again exceeds subsidence a new regressive cycle starts."

The author is very grateful to Dr. Ginsburg for supplying some additional information that is not obvious from the abstract. The subsidence must be differential and a broad, open shelf that gradually tilts seaward is visualized. All of the sediment produced on the shelf is transported shoreward, where it accumulates as a wedge-shaped deposit that builds into a tidal bank.

A further analysis of the model is interesting for two reasons. First, the model has been and still is seriously suggested as a possible mechanism to explain cyclicity on carbonate platforms (see Goldhammer *et al.*, 1987, for references). It also has been quoted frequently, almost as the prototype of "autocyclicity." Secondly, the problem is eminently suitable for computer simulation; with very simple models, considerable insight into the mechanics of the processes can be gained.

Important terminological problems are linked to the first point, such as: "What is an autocycle?" and "Is the Ginsburg process one which is capable of producing self-regulated cycles?" Linked to the second point are such questions as: "At what point does it become more successful to use computer modeling rather than simple pencil and paper?" "How closely must a model simulate real processes in nature before it becomes an acceptable explanation of such processes?" Or, more generally, "How useful are computer simulations of sedimentary processes?"

A Simplified Model

The partial answer to the last question above must be that, in order to prepare a problem for computer simulation, it has to be sharply defined and this can only be of benefit. Although the tilting shelf as proposed by Ginsburg is attractive from the geological point of view, a slightly simpler model will be considered first.

It is assumed that a horizontal shelf of width L is continuously and constantly sinking with rate k. The shoreline is kept fixed and all sediment is immediately transported towards the shore, where it builds a tidal bench which has its surface at sea level. This sediment will be called the *bench* for simplicity and its width ℓ increases gradually (Fig. 1). In contrast, the shelf keeps its width, but the bench-free area gradually decreases. Sediment production is strictly proportional to the bench-free area, which is the area below sea level. One can therefore write:

$$s(L - \ell)dt = \ell dw + wd\ell \tag{1}$$

where s is the constant factor which determines sediment production and w is the water depth. By setting $w = kt + a$, that is, assuming constant

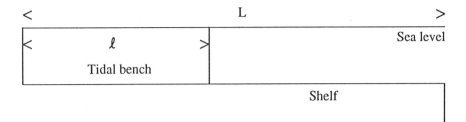

Figure 1: The horizontal shelf model.

subsidence rates, one can write:

$$(a + k)\frac{d\ell}{dt} + \ell(k + s) - sL = 0. \tag{2}$$

The solution is:

$$\ell = \frac{sL}{(k + s)}\left(1 - \left(\frac{a + kt}{a}\right)^{\frac{k+s}{s}}\right). \tag{3}$$

It is assumed that at time $t = 0$, ℓ equals zero and a corresponds to the depth of the empty shelf. If t becomes large, the limit of the tidal bench will be approached with the width:

$$\ell = sL/(k + s). \tag{4}$$

As long as the subsidence rate is greater than zero, the bench margin will never reach the shelf edge; and since the process is a continuous one, no cyclicity will be generated. This is not surprising, because the system lacks the essential link or feedback between the driving force—tectonic subsidence and sedimentation. According to the model, sedimentation clearly has no influence on subsidence and as long as sedimentation is independent of water depth, which is in itself determined by subsidence, the reverse is also true. The main fallacy of the Ginsburg model is the assumption that a new transgressive cycle begins. There is no reason for it to do so.

Any of these conclusions are easily reached without computer modeling, but it is not difficult to write an algorithm which performs the functions of this model (see Fig. 2). It should be noted, however, that care must be taken when very large values of sediment production are investigated together with very low subsidence rates. This is a situation in which the bench margin comes close to the shelf edge. In such a case the model may become unstable, with the result that it begins to oscillate. Such misleading behavior can be avoided by making the time steps sufficiently small.

True oscillations can be generated by introducing a functional relationship between water depth, and therefore indirectly, subsidence and sedimentation. This can be achieved by assuming a critical water depth Wc, below

Sea level

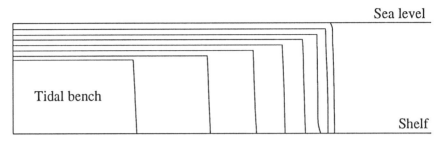

Tidal bench

Shelf

Figure 2: The consecutive filling of a constantly sinking shelf. The shore is to the left and the shelf edge to the right.

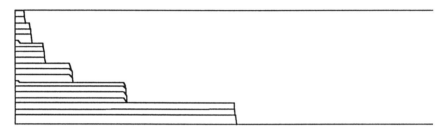

Figure 3: Computer simulation of a horizontal sinking shelf. A critical depth prevents sedimentation and causes stepwise retreat of the tidal bench.

which sediment production stops. A minimum water depth is implied by the constant a in Eq. (3).

The introduction of a maximum depth is a reasonable assumption which could be interpreted as approximating declining production that may be coupled to declining light penetration. Sedimentation now becomes cyclic in the sense that once the critical depth is reached, the sediment-producing area is suddenly reduced and, from this moment, the width of the tidal bench is reduced. The geometry of the process is quite clear: the first cycle will have a width $sL/(s-k)$ and the sediment production for the next cycle will be proportional to this. After n cycles, the width of the bench will have been reduced to $sL/(s+k)$. The cycle thickness in this case is constant and is equal to Wc/k (Figs. 3 and 4).

The cyclicity produced by this model is only recognizable by the position of the bench margin. If sedimentation were to be monitored inside the tidal bench without knowing the position of the margin, no changes in sedimentation would be recorded. The vertical sedimentation rates are constant and show no cyclic variation.

One advantage of computer simulation is that models can be tested for stability. Either the initial conditions or the parameters during a run are randomly disturbed and the effect of this is examined. The horizontal shelf

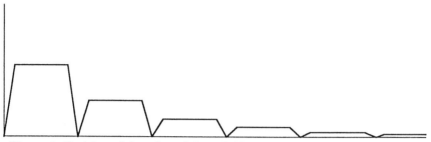

Figure 4: Position of the edge of the tidal bench in the successive time steps seen in Figure 3.

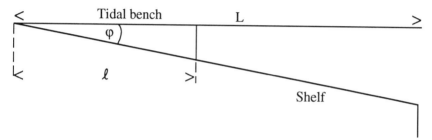

Figure 5: The inclined shelf model.

model is found to be relatively insensitive to small random departures from the horizontal surface of the shelf. However, it is highly sensitive to any systematic change of water depth in the sediment-producing shelf. For example, a small inclination can lead to practically unpredictable behavior. The same is true if subsidence rates are changed by random fluctuations.

The Inclined Shelf Model

As mentioned in the discussion of the Ginsburg model, Ginsburg assumes that differential subsidence is an essential feature of his model. This is only true insofar as it could explain a fixed shoreline and in this instance it is more realistic than the horizontal shelf model. The two models are, however, quite similar in many respects.

A shelf of width L rotates at a constant angular rate ω. Sediment is accumulated in a wedge formed by the shelf with inclination $\varphi = (\omega t + \omega_o)$ and a width 1 (Fig. 5). ω_o is the inclination at time $=$ zero. The area of the wedge is proportional to the width of the shelf which is not covered by the tidal bank:

$$(L - \ell)s \cos\varphi = \frac{1}{2}\ell \sin 2\varphi \frac{d\ell}{dt} + \frac{1}{2}\ell^2 \cos 2\varphi \frac{d\varphi}{dt}. \tag{5}$$

Assuming constant rotation, one can write:

$$\frac{d\ell}{dt}\ell\varphi + \frac{1}{2}\ell^2\omega = s(L - \ell)$$

$$(\omega_o + \omega t)\frac{d\ell}{dt} = \frac{sL}{\ell} - \frac{\ell\omega}{2} - s.$$

Integration of the equation gives:

$$-\frac{1}{\omega}\ln\frac{\omega_o + \omega t}{\omega_o} = \frac{1}{2a}\ln\frac{a + \ell + s/\omega}{a - \ell - s/\omega}; \quad a^2 = \left(\frac{s}{\omega}\right)^2 - \frac{2sL}{\omega}$$

and from this one obtains:

$$\ell = \left(\frac{1}{\omega t}\right)^{\frac{2a}{\omega}}(a - s\omega) - (a + s\omega).$$

When t becomes large, the first term disappears and the equilibrium width of the shelf becomes:

$$\ell = -\frac{s}{\omega} + \sqrt{\left(\frac{s}{\omega}\right)^2 + \frac{2sL}{\omega}}. \tag{6}$$

Once again, any combination of s and ω will lead to an equilibrium position of the tidal bench. To achieve a cyclic transgression–regression movement in this case, one must introduce W_0 and W_1, two critical water depths. W_1 is a maximum depth that switches off sedimentation and therefore prevents the bench from spreading to its maximum width. The additional minimum depth W_0 is needed to prevent any sediment formation in shallow water. Without this lower critical depth, continuous subsidence would permit continuous sediment production and a stable equilibrium would be established.

With two critical values the geometry of the sediment-producing area becomes very complicated and the problem definitely becomes one more suitable for simulation than for theoretical analysis. It is possible to demonstrate some features common to all experiments and these are illustrated by an actual simulation (Figs. 6 and 7). The plot of bench margin position (Fig. 7) as a function of time shows that the amplitude of the oscillation declines exponentially as in the previous model. However, at the same time, the wavelengths of the oscillations increase. The model, therefore, cannot produce cycles of equal thickness. The geometry also makes it inevitable that successive cycles retreat (Fig. 6) and a progradation of cycles is impossible. Although no information is given by the model as to how sediment is transported to the tidal bench and how it is deposited, the geometry of the system implies that sedimentation rates are higher in the distal part of the bench than near the shore. This effect can be seen in simulations which supply time lines, but this may not be the case in real examples. In the absence of exact timing, cycles which are examined without knowing the position of the bench margins could not be recognized.

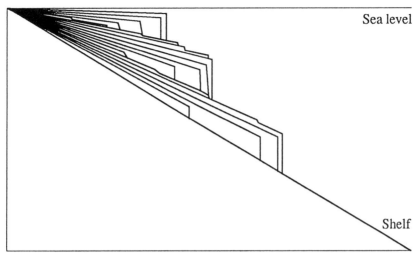

Figure 6: Computer simulation of a constantly tilting shelf. Two limiting water depths prevent sedimentation in deep and shallow water.

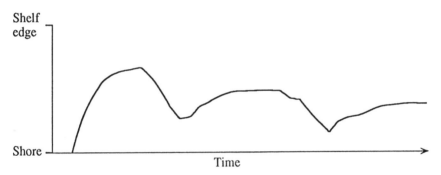

Figure 7: Position of tidal bench margin for the simulation of the tilting shelf in Figure 6.

Discussion

The analysis of the Ginsburg mechanism was primarily undertaken to investigate whether it can be used as an explanation for cyclic sediments. In other words, whether the model provides a self-oscillating system in the terminology of physics (Andronov *et al.*, 1966). Such systems are defined as those which produce oscillations that may or may not be periodic from a non-oscillating source of energy. Unfortunately, geologists use the term "autocycle," which is based on a completely different definition. An autocycle is something that is generated within a certain area. Beerbower (1964), who introduced this term, used the expression within the sedimentary prism, which presumably meant an area and not a volume of sediment.

The area aspect of the system is important and the author has proposed elsewhere (Schwarzacher, 1993) that this should be called the *domain of the cyclic system*. The term autocycle or "autonomous cycle" (Brinkmann, 1932) should be used with the same meaning as in physics, where it describes self-oscillating systems.

Applied to the sedimentation problem, one finds that the driving force of the Ginsburg model is tectonic subsidence; sedimentation is controlled by it. To produce cycles, mechanisms that switch sedimentation on and off are needed. The Ginsburg model provides neither. What it does provide is a mechanism that switches off the progradation of the tidal bench.

Oscillations can be generated by introduction of a critical water depth. It is essential that such a critical water depth be changed by the sedimentation process itself. For example, if sedimentation gradually reduces water to a critical depth which prevents further sedimentation, this may be part of a self-regulating process. In contrast, assume the shelf is continuously sinking and that when a critical depth is reached it will be filled instantaneously with sediment. Sinking continues until once again the critical depth is reached. This process clearly needs an outside mechanism which switches the sediment supply on and off, and this process is not self-regulating.

With the modification of critical depths, the horizontal and the inclined shelf models can produce self-regulated oscillations, but they cannot explain periods in which no sedimentation occurs. As a consequence, the available space for sediment accumulation, as well as sediment generation, is continuously reduced and any tidal bench which develops on the shelf will retreat stepwise. Only the horizontal shelf model can produce cycles of equal thickness. Any deviation from the horizontal has a pronounced effect and soon leads to an unpredictable sediment pattern.

The retreat of the bench, too, is very rapid, particularly in shallow water. For example, a 25-km-wide shelf with a critical water depth of 10 m and sediment production resulting in a 20-km-wide tidal shelf in the first cycle will have shrunk to 8.1 km after 5 cycles and to 2.6 km after 10. To counteract this, one would have to assume an exponential increase in sediment production from the bench edge to the shelf margin. To generate cycles actually recognizable in vertical sections, sediment production would have to change systematically across the shelf. This could be the case if there were a strong zonation of biotopes on the shelf. Considering that the production and zonation of sediment types are likely to be strongly correlated with water depth, the Ginsburg model could not be used to describe such processes.

It would be desirable, therefore, if authors who propose the Ginsburg model as an explanation for regular and often complex cycles were more specific in explaining how the model is supposed to work. Most geologists, including this author, understand that reality is much more complex than simple models and that many models appear not to work in nature. However, a model which does not work in theory is unacceptable.

Acknowledgments

I am grateful to Dr. D.G. Thompson for help in integrating Eq. (5). The paper was discussed in detail with my son, Dr. W. Schwarzacher.

References

Andronov, A.A., Vitt, A.A. and Khaikin, S.E., 1966, *Theory of Oscillators:* Pergamon, Oxford, 815 pp. [English transl.].

Beerbower, I.R., 1964, Cyclothems and cyclic depositional mechanisms in alluvial plain sedimentation: *Kansas Geol. Survey Bull.*, v. 169, no. 1, p. 31–42.

Brinkmann, R., 1932, Über Schichtung und ihre Bedingungen: *Fortschr. Geol. Pal.*, v. 11, 187 pp.

Ginsburg, R.N., 1971, Landward movement of carbonate mud: New model for regressive cycles in carbonates (Abs.): *Am. Assoc. Petroleum Geologists Bull.*, v. 55, no. 2, p. 340.

Goldhammer, R.K., Dunn, P.A. and Hardie, L.A., 1987, High frequency glacio-eustatic sealevel oscillations with Milankovitch characteristics recorded in Middle Triassic platform carbonates in northern Italy: *Am. Jour. Sci.*, v. 287, p. 853–892.

Schwarzacher, W., 1993, Milankovitch cycles in the pre-Pleistocene stratigraphic record, *in* Hailwood, E. and Kidd, R., (eds.): *High Resolution Stratigraphy in Modern and Ancient Marine Sequences*, Geol. Soc. London [in press].

13

QUANTITATIVE GENETICS IN PALEONTOLOGY: EVOLUTION IN TERTIARY OSTRACODA

Richard A. Reyment
and K.G. McKenzie

The applied science of quantitative genetics has a fairly long history in animal husbandry. In its original form, as developed mainly in the hands of D.S. Falconer, there was little obvious connection to the study of evolutionary change in the morphology of fossil species (a summary is given in Falconer, 1981). However, Lande (1976, 1979) derived several generalizations that were aimed at adapting existing theory and method to the study of evolution in the phenotype.

The analysis of shape is a subject that has preoccupied quantitative biologists and biometricians for almost all of our present century. Legion are the methodologies that have been put forward, but the first real solution is quite recent and is due to F.L. Bookstein. A complete account of the history and current state of the art has appeared recently (Bookstein, 1991), and there is a paleontologically oriented summary in Chapter 4 of Reyment (1991). Topics considered in the present brief overview are:

1. Evolution in morphological characters of fossil species related to hypotheses of selection and random genetic drift;

2. Evolution in shape divorced from the effects of size;

3. Erection of a valid hypothesis for the analysis of morphological changes in time.

An important conceptual basis for the application of quantitative genetics to evolutionary biology is that of phenotypic plasticity (Schmalhausen, 1949), i.e., the expression of ecophenotypy. It is a necessary step in an analysis to distinguish between variation arising solely from internal, genetic causes and variation caused by morphological responses to ecological factors.

Evolution in the Phenotype for One Variable

Evolution in the phenotype is most easily approached via consideration of a single distance measure, even though this is not realistic since a meaningful evaluation of an evolutionary sequence on just one character cannot tell us very much. For demonstration purposes we will use a length or distance measured between two diagnostic locations on the organism and study changes by means of models for assessing selection and drift. In the accompanying example we use as our characteristic the height of the carapace of Tertiary ostracodes, that is, the distance between the cardinal angle and the base of the anteroventral rounding.

This procedure, used by Lande (1976), is based on a measure of the rate of evolution first suggested by Haldane (1949), namely, the rate of change of units of the phenotypic standard deviation over time. This measure is easy to work with for mathematical modeling, although it is not the only possibility and requires particular caution.

Model of Selection without Random Genetic Drift

The primary model of Lande (1976) is based on truncation selection, well known from experimental work in animal husbandry, in which a fixed proportion of extreme forms are culled from the population for every generation. All phenotypes have a fitness of 1, except those beyond the truncation point which have a fitness of zero. Charlesworth (1984) gives an alternative model.

Let b denote the number of phenotypic standard deviations from the average phenotype, represented as some character with mean μ_0 and standard deviation σ, to the truncation point, which is located at $\mu_t - b\sigma$. The significance of this step is that it assesses the evolutionary relevance of a morphological shift on the scale of the standard deviations. A practical convention applies here whereby it is assumed that a shift of three standard deviations has evolutionary meaning, as opposed to smaller shifts that could be due to ecological riding and/or random fluctuations.

For weak truncation selection on a normally distributed character and unit average fitness, Lande (1976) showed that the selection response R in one generation is

$$R \simeq \pm \frac{h^2 \sigma}{(2\pi)^{1/2}} \exp\left(\frac{-b^2}{2}\right) \tag{1}$$

and that

$$b = \pm \left[-2\ln(2\pi)^{1/2} \left(\frac{|\mu_t - \mu_0|}{\sigma h^2 t} \right)^{1/2} \right]. \tag{2}$$

The value of b yielded by Eq. (2) is readily converted to an estimate of the proportion of the population per million individuals culled each generation to produce the observed morphological change by consulting the appropriate entry in a table of the standard normal integral.

What is h^2 here? It is the heritability of the character being analyzed. In the normal course of events, this quantity must be estimated experimentally for parent–offspring relationships. In paleontology this is, of course, not possible. Some other way must be tried, and Reyment (1982) proposed a utilitarian solution to the problem by recourse to a nomogram over a range of likely heritabilities (from 0.2 to 0.8). The method is illustrated in Chapter 5 of Reyment (1991).

The Rate of Evolution

The problem posed by estimating the rate of evolution cannot be solved by the simple means suggested in discussions of punctuation–gradualism inconsistency. This popular concept is tainted with fallaciousness because *any* time series composed of morphological observations can adopt a persuasive trend (anagenesis), saltations, or stasis as normal manifestations of a random walk (Bookstein, 1987; Bookstein and Reyment, 1993). In fact, it is only possible to be reasonably categorical about a random walk under conditions in which there is an external variable, *not* time, which can be related to the morphological series or where the net shift in some character, from start to finish of the observed sequence, encompasses at least three standard deviations—*i.e.*, where the span of the morphological change is too great to be solely due to the outcome of the effects of random variation.

There is a situation that is, in part, a fusion of the behavior of the morphology of several species observed simultaneously for variational conformity in the same samples (Reyment, 1971). Estimating the rate of evolution requires considerable attention to mathematical detail. Crow (1986, p. 197), summarizing his views on the subject, underlined the need to realize that evolutionary rates in the same phylogeny are vastly unequal at different times.

Random Genetic Drift and Phenotypic Evolution

We shall now consider random genetic drift in a morphological character in the absence of selection. This is the other extreme of the model considered in the foregoing section, and in real life the situation often lies somewhere in between (*cf.* Crow, 1986, p. 184). The random genetic drift model goes back to Wright (1931), who apparently was first to realize that drift in small populations could be an important factor in evolution. A paleontologically oriented discussion is given by Reyment (1991, p. 180).

Lande (1976, p. 321) derived an approximate statistical test for judging whether an evolutionary event could have arisen in the hypothetical situation of pure random genetic drift, without any selective effects in a population of size N. This procedure estimates how small the effective population size would have to be in order to produce an observed morphological change in the mean by drift alone. Under this assumption, and after t generations, the probability distribution of the average phenotype is normal with variance

$$\text{var}\,\mu_t = h^2\sigma^2 t/N\,. \tag{3}$$

The effective population size N^* at which there is a 5% chance of drifting a distance at least $\mu_t - \mu_0$ in either morphological direction in t generations is obtained when the magnitude of the morphological change is $1.96\,\text{var}\,(\mu_t)^{1/2}$. This leads to the relationship:

$$N^* = (1.96)^2 h^2 t\sigma^2/[\mu_t - \mu_0]^2\,. \tag{4}$$

If $N > N^*$ the shift in means is greater than could have resulted from drift alone. Note that sampling errors in estimates of μ_t and μ_0 are not considered in Lande's *ad hoc* test. There are several problems involved in using these formulae, of which the following are the most serious:

1. Finding a value of the heritability h^2;

2. Reliability of the estimates of t, the number of generations;

3. Combining samples—different localities must not be pooled, because this destroys any chance of judging the regional validity of a purported morphological shift.

Multivariate Analogues

Consider now the vector of k characters $\mathbf{X}^t = (x_1, x_2, \ldots x_k)$ with a hypothetical phenotypic covariance matrix $\mathbf{S}_P$ and genetic covariance matrix $\mathbf{S}_G$. In the model these are considered to remain constant over time. We employ once again a model of truncation selection to account for a change in the population mean vector from μ_0 at generation 0 to μ_t in generation t. This is done by invoking an Index of Selection:

$$I = (\mu_t - \mu_0)^t \mathbf{S}_G^{-1}\mathbf{X} = \sum a_i x_i\,, \tag{5}$$

which is a linear combination of the original characters. Statistically, I represents the linear discriminant function scores, where the a_i are the discriminant coefficients. Index values for generations 0 and t can be calculated and Eq. (5) used to estimate the minimum level of truncation selection needed to "explain" the observed morphological change.

For practical purposes, $\mathbf{S}_G$ and $\mathbf{S}_P$ can be considered to be connected by a constant of proportionality. Cheverud (1988) showed that the genotypic covariance and phenotypic covariance matrices maintain a constant relationship between each other. This is of consequence for the application of the theory of quantitative genetics to evolution in the morphology of fossils. Hence,

$$\mathbf{S}_G \approx h^2_{ave}\mathbf{S}_P$$

where h^2_{ave} denotes an average heritability value.

The estimate of b for obtaining the minimum selective mortality per million individuals per generation is then obtained by computing Eq. (6). It is sometimes useful to examine results for several heritabilities and this can be done most usefully by plotting a set of values of b against a selected suite of heritabilities, say from 0.2 to 0.8. The superscript I denotes the selection index.

$$\hat{b} = \pm \left[\frac{-2\ln(2\pi)^{1/2}\,|\,\mu^I_t - \mu^I_0\,|}{\sigma^I \bar{h}^2 t} \right]^{1/2} \tag{6}$$

The alternative hypothesis of random genetic drift in the absence of selection is approximately "tested" using the chi-squared value

$$\chi^2_p = (\mu_t - \mu_0)'\mathbf{S}_G^{-1}(\mu_t - \mu_0)N/t\,, \tag{7}$$

which corresponds to a weighted Mahalonobis' generalized statistical distance. Setting this statistic equal to the upper 5% point of the χ^2 distribution, say $\chi^2_{0.05}$, and solving for N^* yields the required formula for population size:

$$N^* = t\chi^2_{0.05} / \left[(\mu_t - \mu_0)'\mathbf{S}_G^{-1}(\mu_t - \mu_0) \right]. \tag{8}$$

This is the maximum effective population size that is consistent with the hypothesis of drift in the set of characters in the absence of selection.

Quantitative Genetics of Shape Alone

Until now we have been concerned with contemplating evolution in the phenotype based on measures of size alone or with a vector of measures in which size and shape variability are inextricably confounded. This is not a particularly constructive way to penetrate the problems of evolution in morphometrical relationships; it would be more to the point if some measure of true shape were to be substituted in the foregoing equations. Interesting possibilities are suggested by landmarks, landmarks with outlines, outlines without landmarks, pictures with landmarks, *etc.* (Bookstein, 1991).

Several possibilities can be considered for doing this:

(1) Average "eigenshapes" could be tried. Eigenshapes are a way of designating variation in the outline of an organism, and if this is the extent of one's interest then the method is useful. A critical discussion of this class of techniques is given by Bookstein (1991).

(2) Descriptors in shape-space could be used as the phenotypic parameters and could be inserted into Lande's equations.

(3) The dominant dimensions of shape–space could be found and then used in an analysis by selection gradients using factors rather than raw indicators, for example, the relative warps of Bookstein (1991).

None of this may seem to constitute a real advance over what has been done before in application of the theory of quantitative genetics to evolution in the phenotype. However, one advantage of geometric methods is the possibility of introducing the bending energy matrix in addition to genetic and phenotypic covariances. Hence, application of multivariate quantitative genetics to a variety of species and selection gradients should permit enquiry as to whether certain directions of geometric evolution (*i.e.*, shape) were less accessible than others. In turn, this can yield indications about the mechanisms by which these evolutionary changes are coded. This is a challenging future direction for research.

Methodological Underpinnings

Landmarks are a key concept in geometric morphometrics used to produce shape coordinates. The construction of a flexion relating any configuration of landmarks to any others by a thin-plate spline interpolation permits the extraction of statistically useful components. The use of a spline to deform or warp images of fossils into the geometry of average shape can be related to the time ordering of the material.

Shape coordinates offer the simplest way of obtaining the mean form in which two landmarks are arbitrarily selected as the baseline of a triangle, the apex of which is formed by each of the remaining landmarks in turn. These apical locations of landmarks are plotted with respect to the baseline by rotating, translating, and rescaling all the individual data sets as necessary. See Chapter 4 of Reyment (1991) for several examples of the method.

Now comes a crucial step. The construction of a method to warp from one shape to another requires a flexible, data-oriented function that will *map any subset of the landmarks from one form or from the averaged form onto the same set of landmarks on another form*. This can be done in several ways, but the most accessible from the standpoint of data analysis and quantification of patterns is thin-plate spline interpolation, developed 15 years ago by the French mathematicians Duchon (1976) and Meinguet (1979). Their technique has since spread to computer graphics and other fields. It serves to interpolate local nonlinearities of landmark rearrangement, the generic term for the method being *principal warps* (Bookstein, 1991). Decomposition of the spline yields features of change in shape, the first of which is the affine part. The subsequent decompositions, which are referred to as *partial warps*, are obtained in terms of figurative bending energy, each of which relates to some deformational feature of the evolutionary mapping of the older form into the younger one. A completely worked example for fossil ostracodes is given by Reyment and Bookstein (1993). For a sample of specimens, the method is exactly analogous to principal component analysis of the covariance matrix of shape coordinates. This provides the structure of landmark variability by what is known as the method of relative warps. For our present exposition, the partial warps obtained from the principal warps are germane. Two worked examples are provided in Chapters 4 and 8 of Reyment (1991).

Overview of Main Geometrical Concepts

Let $\mathbf{X}_i$ denote the $(2 \times p)$ data matrix of digitized x–y-coordinates for the ith specimen of a species and let $\mathbf{X}$ be the $(2n \times p)$ matrix of all n specimens of the sample. We shall also write $\mathbf{X}_x$. for the $(n \times p)$ matrix (vector) of x-coordinates alone and $\mathbf{X}_{.y}$ for the corresponding array of y-coordinates. It is required to align the n specimens so that a reference specimen (*i.e.*, an averaged specimen) $\mathbf{X}_e$ can be computed. This can be done by some suitable method of superposition, such as an affine resistant-fit procedure. This is a composite on which *all the essential features of a species, morph, or any other taxonomic unit can be displayed and mathematically specified to any degree desired.*

Bookstein's method of shape coordinates can be applied to a previously defined baseline. The location of each landmark in the reference specimen can be computed as the mean x–y-coordinates across the objects.

The reference specimen is used to define the principal warps. The bending energy matrix $\mathbf{L}_p^{-1}$ is found for the reference specimen by the following sequence of steps restated from Bookstein (1991): Begin by constructing the matrix $\mathbf{L}$

$$\mathbf{L}_p = \left[\begin{array}{cc} \mathbf{P}_k & \mathbf{Q} \\ \mathbf{Q}^t & \mathbf{0} \end{array} \right] \tag{9}$$

where $\mathbf{P}_k$ is a matrix with zeros down its diagonal and with off-diagonal elements defined by

$$U_{(r_{ij})} = r_{ij}^2 \log r_{ij}^2 .$$

Alternatively, $\log r_{ij}$ may be used, which will only influence scaling. The term r_{ij}^2 denotes the square of the distance between landmarks i and j of the reference specimen. The matrix $\mathbf{Q}$ consists of three columns and p rows, the first of which is a unit vector, the second is the vector of x-coordinates, and the third is the vector of y-coordinates.

The upper left submatrix of the inverse of $\mathbf{L}$ is the bending energy matrix. It has latent roots Φ and latent vectors $\mathbf{E}(p \times p)$ in which columns correspond to the normalized latent vectors and rows to the landmarks. These are the principal warps. Because of the way in which $\mathbf{L}$ is constructed (incorporating the lower right matrix of zeros), there can be at most $p - 3$ nonzero latent roots (Rohlf, 1993).

The next step is to assemble the weight matrix $\mathbf{W}$ of order $(n \times 2(p-3))$. This is a scaled projection of the x- and y-coordinates of the deviation of the n specimens from the consensus specimen onto the principal warps having nontrivial latent roots. We shall write this as

$$\mathbf{W} = [\mathbf{W}_x \mid \mathbf{W}_y]$$

where

$$\mathbf{W}_x = n^{-1/2}\mathbf{V}_x\Phi^{\alpha/2}\mathbf{E}^t$$

and $\mathbf{W}_y$ is the analogue for y. Also,

$$\mathbf{V}_x = \mathbf{X}_x - \mathbf{1}_n \otimes [1|0]\mathbf{X}_c$$

and

$$\mathbf{V}_y = \mathbf{X}_y - \mathbf{1}_n \otimes [0|1]\mathbf{X}_c .$$

$\mathbf{1}_n$ is a column vector of n ones and $\mathbf{V}_x$ is an $(n \times p)$ matrix of the x-coordinates of the differences between the n specimens and the reference specimen. $\mathbf{V}_y$ is the equivalent matrix of the y-coordinates (Rohlf, 1933).

The singular value decomposition of the weight matrix yields:

$$\mathbf{W} = \mathbf{PDR}^t, \tag{10}$$

where $\mathbf{P}$ is an $(n \times 2(p - 3))$ matrix having as its columns the contributions of each of the $2(p - 3)$ relative warps to the n specimens. $\mathbf{D}$ is a diagonal matrix of singular values and $\mathbf{R}$ is a $(2(p - 3) \times 2(p - 3))$ matrix whose columns are the relative warps in terms of the principal warps. The first $(p-3)$ rows of $\mathbf{R}$ pertain to the x-coordinates and the remaining $(p-3)$ rows pertain to the y-coordinates. The matrix of relative warps represents the displacements of the positions of the landmarks on the reference specimen that one would expect if an individual differed from the consensus specimen by a unit change in a specified relative warp. The method of relative warps is a direct analogue of principal component analysis.

The application of geometric morphometrics to quantitative genetics, outlined in the introduction, can be made to proceed via the relative warps. The eigenanalysis of the sample covariance matrix of shape coordinates can be taken with respect to the bending energy matrix $\mathbf{L}_p^{-1}$ evaluated at the sample mean shape. Another solution is to confine the analysis to the non-linear part, which is the interesting bit from the aspect of evolutionary theory. One can then use scores of relative warps for each specimen or species and analyze them by multivariate methods. The weight matrix, $\mathbf{W}$, of shape data also can be utilized without going to the principal component decomposition that produces the relative warps.

Table 1: Results of Quantitative Genetic
Computations for $h^2 = 0.4$

Number of selective deaths per million/generation	Population size N^*
0	2,000,000
1	1,000,000
2	750,000

Discussion

The practical implications of the ideas summarized here can best be illustrated by an example. We consider an Australasian lineage of the ostracode genus *Quasibradleya* from the Tertiary. The species are *Quasibradleya momitea* McKenzie, Reyment and Reyment from the middle?–upper Eocene of southeastern Victoria, *Q. prodictyonites* Benson from the lower Oligocene of New Zealand and *Q. paradictyonites* Benson from the Oligo-Miocene of Tasmania (Benson, 1972), *Q. janjukiana* McKenzie, Reyment and Reyment from the upper Oligocene of southeastern Victoria (McKenzie *et al.*, 1991), Australia, and *Q. praemackenziei* Whatley and Downing from the early middle Miocene of southeastern Victoria, Australia (McKenzie and Peypouquet, 1984).

We shall exemplify the quantitative genetic method of analysis by applying it to two species, *Q. paradictyonites* and *Q. janjukiana*, that are closely located in time and place and can reasonably be assumed to be part of the same lineage. They represent a time range of about 1 to 2 million years in the late Oligocene. The results obtained by applying Eqs. (6) and (7) to a decrease in height are listed in Table 1. This table gives computed values for three feasible numbers of generations. These results are very informative. They show that if the observed change could have occurred by selection on its own, such selection would have been very weak indeed. However, this is not a very likely model. If we examine the column for minimum population sizes in the absence of selection, the results provide the basis of a workable hypothesis. Recalling that the value of N^* refers to viable individuals, then the estimate of minimum population size can be reduced to possibly about 25,000 for one million generations, a value that is not inconsistent with the observed change having been caused by random genetic drift.

We examine now the shape differences between the four species by means of principal warps (Table 2). Nine landmarks were selected. These are: 1) position of the eye tubercle, 2) tip of the posterodorsal spine, 3) mid-spine of the posterior margin, 4) posteroventral spine, 5) posterior termination of the ventrolateral rib, 6) maximum ventral convexity, 7) a feature of the adductor

Table 2: X- and Y-Coordinates for Total
Principal Warps between Species

	X	Y	Bending energy
Q. janjukiana to *Q. praemackenziei*			
	X	Y	Bending energy
1	−0.0294	−0.0593	0.0420
2	−0.0136	−0.0177	0.0035
3	−0.0006	0.0199	0.0009
4	0.0141	−0.0223	0.0013
5	0.0007	0.0852	0.0032
6	−0.0440	0.0328	0.0022
			Total 0.0531
Percent anisotropy = 1.45			
Q. prodictyonites to *Q. janjukiana*			
	X	Y	Bending energy
1	−0.0270	−0.1147	0.1332
2	−0.0057	−0.1147	0.0004
3	−0.0404	0.0416	0.0079
4	−0.0841	−0.0273	0.0145
5	0.0822	−0.0071	0.0030
6	0.0682	−0.0102	0.0035
			Total 0.1625
Percent anisotropy = 7.8			
Q. momitea to *Q. praemackenziei*			
	X	Y	Bending energy
1	−0.0460	−0.0741	0.0076
2	−0.0422	0.1420	0.0219
3	−0.0317	−0.0268	0.0017
4	−0.0327	−0.0622	0.0049
5	0.1005	−0.0716	0.0152
6	0.1353	−0.0352	0.0196
			Total 0.3766
Percent anisotropy = 49.02			

muscle tubercle, 8) center of ventrolateral rounding of the anterior rib, and 9) center of rounding of the anterior margin. Geometric morphometric analysis discloses that the farther apart in time the species are located, the greater are the differences in shape. Size characters considered one at a time are not very different, but when divergences in landmark locations are assessed, the effects of evolutionary differentiation become readily apparent. The apportionment between changes in affine shape and non-affine shape are variable. The greatest anisotropy is found for the comparison between the middle? Eocene and middle Miocene members of the lineage. This is 49%, which shows that a considerable part of the morphological differentiation lies in a uniform change in the locations of the nine landmarks.

We begin by examining results for the Eocene–Miocene pair of species. We note that these two species inhabited similar environments; therefore, the observed changes in shape are less likely to be wholly ecophenotypic. The second partial warp is the strongest of the six, as is shown by the appropriate x–y-coordinates in Table 2. (There can be only $N - 3$ partial warps, where N is the number of landmarks.)

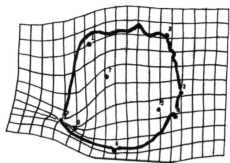

Figure 1: Second partial warp for the transformation from *Quasibradleya momitea* (middle?–late Eocene) to *Q. praemackenziei* (middle Miocene). Bending energy = 0.0219. There is global lateral compression and strong anteroventral squeezing.

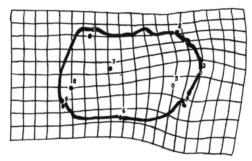

Figure 2: First partial warp for the transformation from *Quasibradleya momitea* (middle?–late Eocene) to *Q. praemackenziei* (middle Miocene). Bending energy = 0.0076.

Figure 1 illustrates the second partial warp, showing that landmarks 8 and 9 are particularly affected. The fact that the corresponding part of the bending energy is the largest of six values in the bending energy column of Table 2 indicates this is a global deformation. For comparison, Figure 2 shows the first partial warp corresponding to the third smallest bending energy. This means that it represents a small-scale deformational feature—a dilation of the posterior margin mainly affecting landmarks 3, 4, and 5.

By Oligocene time the tendency toward alteration of the anteroventral configuration of the shell seems to have terminated. Figure 3 illustrates the total non-affine warping from early Oligocene to middle Miocene. There is still dilation of the posterior margin, but now the posteroventral region is affected more strongly. The first partial warp, in Figure 4, is connected to the largest fraction of the bending energy (Table 2) and expresses a global deformation in the form of dilation of the posterior margin. It reflects shape

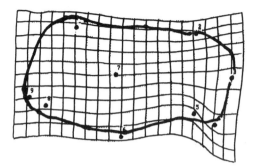

Figure 3: Nonaffine change of landmark configuration for the pair of thin-plate splines for *Q. janjukiana* (late Oligocene) to *Q. praemackenziei* (middle Miocene). Bending energy = 0.0532.

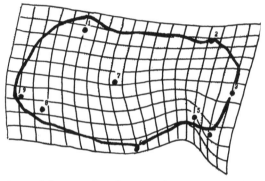

Figure 4: First partial warp for the transformation of *Q. janjukiana* (late Oligocene) to *Q. praemackenziei* (middle Miocene). Bending energy = 0.0420.

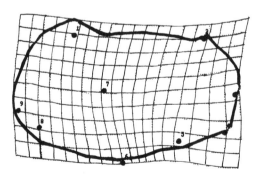

Figure 5: Nonaffine change of landmark configuration for *Q. prodictyonites* (early Oligocene) to *Q. janjukiana* (late Oligocene). Bending energy = 0.1625. There is mediolateral compression and posteroventral dilation.

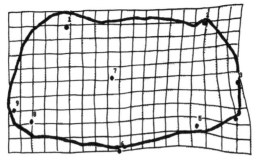

Figure 6: First partial warp for the transformation of *Q. prodictyonites* (early Oligocene) to *Q. janjukiana* (late Oligocene). Bending energy = 0.1625.

change in the posterior part of the shell and involves mainly landmarks 3, 4, and 5. The reduced strength of change in shape is clearly manifested in Figure 6, the sketch of the first partial warp for the transformation from the early Oligocene member of the lineage to the late Oligocene species.

These are the most striking features of our analysis. Shape changes occur throughout the entire lineage, but these are less striking and not unexpected because they comply with the pattern of change indicated by the middle? Eocene and early Oligocene points of entry. It is hard to escape the conclusion that in this admittedly cursory analysis we are concerned with complex change, probably under selective control. There was an important morphometrical change from the Eocene to Oligocene when the site of differentiation in shape appears to have switched from the anteroventral region to the posterior. Thereafter, at least up to the middle Miocene, differentiation remained steadily centered around the posterior landmark constellation. An example of this is provided by the nonaffine warp for early to late Oligocene in Figure 5 and the corresponding first partial warp indicated in Figure 6. A case could probably be made that this represents the effects of genetic drift assisted by selection. Such drift could possibly have been responsible for the reversal in shape separation in the late Eocene.

In a functional sense, the anteroventral differentiation in *Quasibradleya* from middle? Eocene to middle Miocene can be related plausibly to some environmentally cued shift in feeding behavior, since the mouth is located near this site. The posterior dilation in early Oligocene to middle Miocene taxa, on the other hand, is more likely linked to variations in ontogeny with time—since the ecdysistic development of Ostracoda from first instar to adult proceeds by sequential addition of limbs and reproductive systems in the posterior region until achieving maturity. This particular change over time, though more subtle, is more likely to represent a genuine genetic shift.

Acknowledgments

We thank Professors Fred L. Bookstein, Leslie F. Marcus and F. James Rohlf for constructive criticism and the last-named for software used in the geometric computations. The research was supported by grants from the Swedish Natural Science Research Council, including a visiting professorship for K.G. McKenzie.

References

Benson, R.H., 1972, The *Bradleya* problem, with descriptions of two new psychospheric ostracode genera, *Agrenocythere* and *Poseidonamicus* (Ostracoda, Crustacea): *Smithsonian Contributions in Paleobiology,* v. 12, p. 1–138.

Bookstein, F.L., 1987, Random walk and the existence of evolutionary rates: *Paleobiology,* v. 13, p. 446–464.

Bookstein, F.L., 1991, *Morphometric Tools for Landmark Data:* Cambridge University Press, New York, 435 pp.

Bookstein, F.L. and Reyment, R.A., 1992, Random walk and quantitative stratigraphical sequences: *Terra Nova,* v. 4, no. 10.2, p. 147–151.

Charlesworth, B., 1984, Some quantitative methods for studying evolutionary patterns in single characters: *Paleobiology,* v. 10, p. 308–313.

Cheverud, J.M., 1988, A comparison of genotypic and phenotypic correlations: *Evolution,* v. 42, p. 958–968.

Crow, J.F., 1986, *Basic Concepts in Population, Quantitative and Evolutionary Genetics:* Freeman and Co., New York, 273 pp.

Duchon, J., 1976, Interpolation des functions de deux variables suivant la principe de la flexion des plaques minces: *RAIRO Analyse Numérique,* v. 10, p. 5–12.

Falconer, D.S., 1981, *Introduction to Quantitative Genetics, 2nd Ed.:* Longmans Inc., New York, 340 pp.

Haldane, J.B.S., 1949, Suggestions as to quantitative measurement of rates of evolution: *Evolution,* v. 3, p. 51–56.

Lande, R., 1976, Natural selection and random genetic drift in phenotypic evolution: *Evolution,* v. 30, p. 314–334.

Lande, R., 1979, Quantitative genetic analysis of multivariate evolution, applied to brain : body size allometry: *Evolution,* v. 33, p. 402–416.

McKenzie, K.G. and Peypouquet, J.-P., 1984, Oceanic paleoenvironment of the Miocene Fyansford Formation from Fossil Beach, near Mornington, Victoria: *Alcheringa,* v. 8, p. 291–303.

McKenzie, K.G., Reyment, R.A. and Reyment, E.R., 1991, Eocene–Oligocene Ostracoda from South Australia and Victoria, Australia: *Revista Española de Paleontología,* v. 6, p. 135–175.

Meinguet, J., 1979, Multivariate interpolation at arbitrary points made simple: *Zeitschr. für angewandte Mathematik u. Physik,* v. 30, p. 292–304.

Reyment, R.A., 1971, Spectral breakdown of morphometric chronoclines: *Math. Geol.,* v. 2, p. 365–376.

Reyment, R.A., 1982, Phenotypic evolution in a Cretaceous foraminifer: *Evolution,* v. 36, p. 1182–1199.

Reyment, R.A., 1991, *Multidimensional Paleobiology:* Pergamon Press, Oxford, 377 pp. [plus 39-page Appendix by L.F. Marcus].

Reyment, R.A. and Bookstein, F.L., 1993, Infraspecific variability in shape in *Neobuntonia airella:* An exposition of geometric morphometry: *Proceedings 11th International Symposium on Ostracoda,* Balkema, Rotterdam [in press].

Rohlf, F.J., 1993, Relative warp analysis and an example of its application to mosquito wings, *in* Marcus, L.F., *et al.,* (eds.), *Mus. Nac. de Ciencias Naturales:* v. 8, Madrid [in press].

Schmalhausen, I.I., 1949, *Factors of Evolution: The Theory of Stabilizing Selection:* Blakiston, Philadelphia, 327 pp.

Wright, S., 1931, Evolution in Mendelian populations: *Genetics,* v. 16, p. 97–159.

14

AN INTEGRATED APPROACH TO FORWARD MODELING CARBONATE PLATFORM DEVELOPMENT

Helmut Mayer

The forward model presented here is designed to simulate stratigraphic and geometric development of carbonate platforms. Starting from an initial basement geometry, the effects of a number of key variables on water depth are combined for each time increment. This procedure is repeated in an iterative fashion for subsequent time steps. The variables considered include subsidence, carbonate production, sediment redistribution, compaction, isostatic compensation, and eustatic sea-level change. Time- or depth-dependent functions are developed for these variables. Free parameters in these functions allow fitting to realistic magnitudes. A sample simulation demonstrates the characteristics of the model and indicates its usefulness in case studies and predictions.

Modeling Sedimentary Systems

In recent years a number of studies on the modeling of sediment accumulation in various basin settings has been published. Most of them are concerned with clastic basin fill or do not discriminate lithologies (*e.g.*, Turcotte and Kenyon, 1984; Kenyon and Turcotte, 1985; Tetzlaff, 1986; Bitzer and Harbaugh, 1987; Flemings and Jordan, 1987, 1989; Tetzlaff and Harbaugh, 1989; Jervey, 1989), while only few focus on mixed clastic/carbonate systems (*e.g.*, Aigner *et al.*, 1989; Lawrence *et al.*, 1990) or carbonate platforms (*e.g.*, Lerche *et al.*, 1987; Bice, 1988; Demicco and Spencer, 1989; Scaturo *et al.*, 1989).

Sediment accumulation and distribution on a carbonate platform and the adjacent slope represent a highly complex system of numerous interdependent factors which in concert determine the development of the stratigraphy and geometry of the platform. The goal of this study is to develop a model that yields a "best compromise" between two principal targets: representation of all important variables in geologically reasonable functional relationships on the one hand, and simplicity on the other. Forward modeling of sedimentary systems serves to simulate the stratigraphic and geometric evolution of the system, dependent on variations in the input parameters. The purpose of this approach is to establish the critical variables and parameters which dominate the system and to produce a geologically reasonable generic stratigraphic pattern. The next step then would be to use the model to reproduce known patterns of actual modern or ancient sedimentary systems (inverse modeling). This is done by adjustments of the input parameters to achieve a closest fit between model and target. There are two main reasons why sedimentary modeling is more than an exercise. Assuming the model contains all the major variables, we learn about their relative importance and contribution in particular cases. The other reason is that quantitative models allow much better constrained predictions about poorly or partly known systems than would qualitative empirical geological knowledge.

The simulations of Lerche et al. (1987) explored the influence of several variables (food supply, sunlight, temperature, salinity, oxygen content, sea-level change, and subsidence) on the geometric development of a carbonate platform. The authors present seven functions for platform height at a point and instant depending on the respective variable. The paper gives a series of simulations in which the parameters of the functions are changed one by one. Lerche et al. (1987) did not study the interdependences of their variables. Bice (1988) presented a forward model incorporating subsidence, carbonate production and redistribution, and sea-level change without giving the algorithms used. The paper investigates mainly the influence of eustatic sea-level change on the system. Demicco and Spencer (1989) published a computer model that simulates cyclical sedimentation in the near-shore area of a carbonate platform depending on variations in sea level and tidal range. Scaturo et al. (1989) presented an inverse modeling study in which they tried to reproduce the architecture of one particular carbonate platform using a model based on varying accumulation rates, eustatic sea-level fluctuations, and tectonic subsidence. The present paper describes an integrated approach to forward modeling of the stratigraphic and geometric development of carbonate platforms. A quantitative model is developed consisting of six variables which together determine sedimentation on the platform as well as lateral growth patterns. In an iterative procedure the depth below sea level is repeatedly calculated for points along a profile across a shelf area. The model emphasizes the interdependences between variables and stresses their combined effect on platform evolution. A low-resolution example of a simulation is presented. This paper is a short presentation of the approach rather than an exhaustive documentation of detailed simulation results.

Model Setup

The model developed here considers six variables which determine the stratigraphic and geometric evolution of a carbonate platform. These are: subsidence, carbonate production, sediment redistribution, compaction, isostatic compensation, and eustatic sea-level change. Most of these variables are modeled by functions dependent on time, depth, or location. The functional forms chosen are based mainly on empirical observations. These functions contain free parameters to allow for fitting to a reasonable magnitude and range of values in order to obtain geometries which are acceptable on geological grounds. Besides the variables, three initial conditions have to be specified: initial basement geometry, sea level, and shoreline. The model is two dimensional, corresponding to a standard geological cross-section across a shelf perpendicular to the coastline. Individual variables, functions, and parameters will be introduced and discussed below.

Subsidence

The general environment in which a carbonate platform builds up is the marginal area of a subsiding basin. In particular, we consider a passive continental margin setting where carbonate platforms commonly occur. Although other possibilities have been considered (*e.g.*, Sloss, 1982), thermal subsidence due to thinning and cooling of the stretched lithosphere is generally accepted as the basic underlying subsidence mechanism for large sedimentary basins (*e.g.*, McKenzie, 1978; Steckler and Watts, 1978; Sclater and Christie, 1980). Thermal subsidence basically is a function relating depth to the square root of time for the first 60 to 80 m.y. after initiation (Parsons and Sclater, 1977). In real basins, punctuated "tectonic" subsidence events are superimposed onto the thermal trend. Tectonic subsidence will not be considered here, however, because it is too specific to be generalized. Instead, subsidence is expressed as a functional dependence of depth to basement d upon time after initiation t:

$$d = d(t) = c\sqrt{t} + a, \tag{1}$$

where c is a constant scaling factor and a is a constant offset corresponding to the initial depth (Fig. 1). Depth is in kilometers and time in millions of years. In consequence, the subsidence rate decreases with time according to:

$$\frac{d(t)}{dt} = \frac{1}{2}\, ct^{-1/2}. \tag{2}$$

Carbonate Production

For carbonate production, an extremely simplified approach is used. Although several factors that influence carbonate production can be distinguished, such as food supply, light intensity, water temperature, salinity,

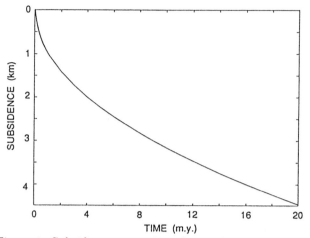

Figure 1: Subsidence curve: Depth as a function of time.

and oxygen content (*e.g.*, Lerche *et al.*, 1987), an integrated empirical carbonate production term as a function of water depth alone seemed to be sufficient and appropriate for this study (*cf.* Bice, 1988). For depths of 0 m to 25 m a high constant production rate has been chosen, while the production rate for greater depths decreases exponentially with depth until there is no production below 300 m:

$$p = \begin{cases} f p_{max} t & \text{for } d \leq 25 \text{ m} \\ e^{-gd} p_{max} t & \text{for } 25m < d \leq 300 \text{ m} , \\ 0 & \text{for } d > 300 \text{ m} \end{cases} \qquad (3)$$

where p is sediment production in meters of vertical sediment column, d is depth below sea surface in meters, f and g are constant scaling factors, p_{max} is the maximum production rate, and t is time (Fig. 2). The parameters p_{max}, f, and g are chosen to guarantee excess sediment production on the shallow platform even at times of rapid subsidence, a feature which is apparent on many platforms in the geologic record (*cf.* Wilson, 1975, p. 16).

Sediment Redistribution

The excess sediment produced on the shallow platform has to be transported and redeposited seaward (progradation), because the platform cannot build up (aggrade) above sea level. Redeposition in the present model is a geometric function of the cross-sectional area of the excess sediment. The amount of carbonate produced on the platform which would grow above sea level (*i.e.*, the excess sediment) is cut off and moved to the available accommodation area which is filled progressively in the seaward direction. At the end of this procedure a free slope of 30° is maintained, as this is the

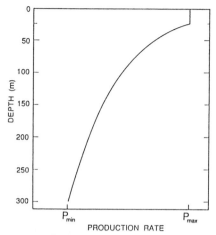

Figure 2: Carbonate production rate as a function of water depth. p_{max} and p_{min} are maximum and minimum rates, respectively.

average for steepest stable slopes in loose carbonate sediment (*cf.* Kenter, 1991). This same function is applied to redeposit sediment eroded from the shallow platform if a rapid sea-level drop causes subaerial exposure.

Compaction

Progressive compaction of sediment with burial can be a substantial factor affecting the geometry of the depositional realm, depending on the type of sediment. Decrease in porosity ϕ with burial depth z follows the law:

$$\phi = \phi_0 e^{-cz}, \tag{4}$$

where ϕ_0 is the compactible initial pore volume, and c is an empirical constant characteristic of individual lithologies (after Athy, 1930, p. 13–14) (Fig. 3). Because the platform setting favors early cementation, the compactible initial porosity in carbonates amounts to only 5–10% (Wildi *et al.*, 1989). For c we adopt a value of $3 * 10^{-4} m^{-1}$ from Wildi *et al.* (1989, p. 825). Taking an upper bound of $\phi_0 = 10\%$ still gives us only a very small correction for compaction at the scale of our model, but it is included here to demonstrate its effect and to keep the model open for higher resolution studies.

Isostatic Compensation

Sediment load causes an isostatic adjustment of the underlying lithosphere. In the present model a local-loading (Airy-type) function is implemented (*cf.* Steckler and Watts, 1978) (Fig. 4). Using the notation as defined in Figure 4, we have

$$w_1 \rho_w + c\rho_c + m_1 \rho_m = w_2 \rho_w + s\rho_s + c\rho_c \tag{5}$$

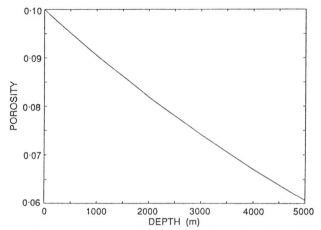

Figure 3: Compaction curve: Porosity as a function of burial depth.

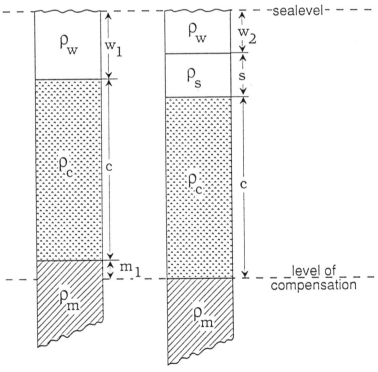

Figure 4: Isostatic compensation in a local-loading (Airy-type) model. Column on right shows adjustment after sediment loading relative to unloaded column on left. ρ_w, ρ_s, ρ_c, and ρ_m are bulk densities of seawater, sediment, crust, and mantle, whose thicknesses are w, s, c, and m. Subscripts 1 and 2 indicate thickness before and after sediment loading.

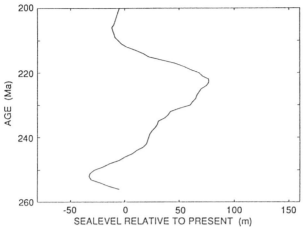

Figure 5: Eustatic sea-level change in geologic history. Long-term global sea-level curve from late Permian to earliest Jurassic (after Haq *et al.*, 1988).

$$w_1 + c + m_1 = w_2 + s + c \tag{6}$$

$$m_1 = s + w_2 - w_1. \tag{7}$$

Combining equations (5) and (7) and rearranging terms yields

$$w_2 = w_1 + s\frac{\rho_m - \rho_s}{\rho_w - \rho_m}. \tag{8}$$

Using approximate values for the densities involved, $\rho_w = 1gcm^{-3}$, $\rho_s = 2gcm^{-3}$, $\rho_m = 3gcm^{-3}$,

$$w_2 = w_1 - \frac{1}{2}\,s. \tag{9}$$

With this simple formula we can compute the isostatic compensation after sediment loading at each point of the grid along the profile.

Eustatic Sea-level Change

The last variable incorporated in the sedimentation model is eustatic sea-level change. Rather than using an abstract sinusoidal sea-level curve, we digitized the widely used long-term sea-level curve published by Haq *et al.* (1987, 1988) and choose appropriate intervals for individual simulations to get closest to the functional character of "real" sea-level changes (Fig. 5).

Modeling Procedure

In the model, water depth is calculated repeatedly at a series of grid points along a profile. Each iteration comprises the consecutive calculation of the effects of each of the variables on water depth. The effects of all variables are added up for each time increment in the order: subsidence, carbonate production, sediment redistribution, compaction, isostatic compensation, and eustatic sea-level change. The depth output of the previous variable function provides the input depth for the following variable function, so the order in which the variables are considered has a strong influence on the result. The particular order was chosen for the following reasons: First, some accommodation space must be available for sediment accumulation. This is provided by subsidence of the basement. Then carbonate sediment must be produced before it can be redistributed. Sediment loading induces compaction and isostatic compensation. The order of the last two variables is based on the longer response time of the lithosphere. There is no obvious reason for eustatic sea-level change being the final variable. One could argue for other positions in the procedure. However, because eustatic sea-level change is unrelated to the other variables, its effect is superimposed at the end. Where no carbonate sediment is deposited, a small amount of pelagic clay is added at a low sedimentation rate of 10 mm/1000 yr. Then time is advanced by a constant increment, and a new iteration is performed. In this way a synthetic stratigraphic succession is produced which simulates the evolution of a carbonate platform.

Example

To demonstrate the characteristics of the model, an example simulation is presented in Figure 6. This simulation uses initial parameters that were determined to be geologically reasonable first-order estimates. The initial basement geometry represents a wide shelf dipping gently seaward (from 0 m to 100 m water depth over a distance of 100 km), which drops across a steep slope into a deep basin (Fig. 6a). For the subsidence function the following parameters were used: c is 1 for depths given in kilometers and times in millions of years. For a, the initial depth at each grid point along the profile is used. In the carbonate production function the parameters are: $f = 1$, $g = 0.005$, and $p_{max} = 150$ m/m.y., which is an intermediate value deduced from the range of production rates calculated for carbonate platforms in the geologic record (*cf.* Schlager, 1981). In the sediment redistribution function a slope angle of 90° rather than 30° is maintained for simplicity. At the large scale of this example this inaccuracy is negligible, while the geometric procedure is made much easier. The effects of compaction and isostatic compensation are added as described above. Eustatic sea-level change for the example corresponds to the rise in the early Middle Triassic (between 240 Ma and 237 Ma) in the curve of Haq *et al.* (1988) (see Fig. 5). Running the model through several time steps reveals a pattern that is similar to that of many ancient carbonate platforms (Fig. 6a–d): The platform builds

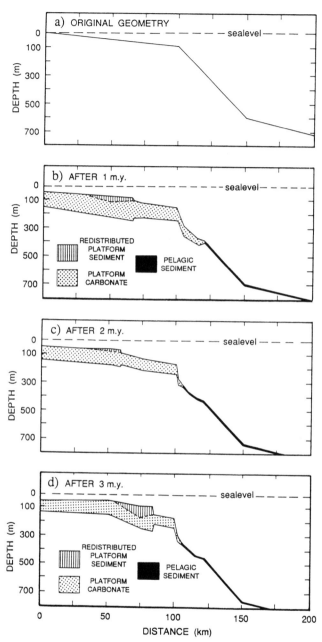

Figure 6: Example simulation; see text for explanation. a) Original basement geometry. b) Platform geometry after first time increment (1 m.y.). c) Platform geometry after second time increment (2 m.y.). d) Platform geometry after third time increment (3 m.y.).

up and keeps up even with rapid subsidence/sea-level rise. There is enough excess sediment produced on the platform to allow for more or less extensive progradation, creating a typical platform geometry. The sharp kinks in the time lines across the platform are artefacts of the low resolution in time and space, combined with the crude local-loading isostatic model used. However, the abruptness of these depth changes should not distract from the fact that they represent the trend. A smoother appearance could be achieved by using smaller time increments and more closely spaced grid points for calculation. Further improvement of the model is possible by implementing a flexural-loading model for isostatic compensation.

Conclusions and Outlook

A simple forward model for carbonate deposition can be based on functional time or depth dependences in a collection of dominating variables. All of the functional relationships as well as the selection of the variables considered represent severe simplifications. Further simplification is introduced by a computation mechanism involving consecutive summation of the effects of changes in individual variables. A high degree of interdependence between variables is retained, but the simultaneity of processes is lost, so that the model is quite sensitive to the length of time steps. However, the model produces results which are in good agreement with known patterns of carbonate platform development. This is achieved without any fine-tuning of the parameters. These properties of the model show its potential in the study of particular carbonate platforms—ancient or recent.

Possible applications of this forward model for carbonate deposition are:

- Using it as a starting point for inverse modeling to learn about the magnitude of contributing parameters for particular platforms.

- Extrapolating or interpolating from better known areas to platform areas which are less well known.

- Predicting the distribution of sedimentary facies in a platform setting for potential use in hydrocarbon exploration.

Acknowledgments

Part of this work was presented at the Dolomieu Conference on Carbonate Platforms and Dolomitization in St. Ulrich, South Tyrol, Italy, in September 1991. The financial support of the organizing committee for participation in the conference is gratefully acknowledged. I also thank Jo Griffith, Scripps Institution of Oceanography, La Jolla, California, for the final drafting of the figures.

References

Aigner, T., Doyle, M., Lawrence, D., Epting, M. and van Vliet, A., 1989, Quantitative modeling of carbonate platforms: Some examples, *in* Crevello, P.D., Wilson, J.L., Sarg, J.F. and Read, J.F., (eds.), *Controls on Carbonate Platform and Basin Development:* Soc. Econ. Paleontologists and Mineralogists Spec. Pub. no. 44, p. 27–37.

Athy, L.F., 1930, Density, porosity, and compaction of sedimentary rocks: *Bulletin Am. Assoc. Petroleum Geologists,* v. 14, no. 1, p. 1–24.

Bice, D., 1988, Synthetic stratigraphy of carbonate platform and basin systems: *Geology,* v. 16, no. 8, p. 703–706.

Bitzer, K. and Harbaugh, J.W., 1987, DEPOSIM: A Macintosh computer model for two-dimensional simulation of transport, deposition, erosion, and compaction of clastic sediments: *Computers & Geosci.,* v. 13, no. 6, p. 611–637.

Demicco, R.V. and Spencer, R.J., 1989, MAPS—A BASIC program to model accumulation of platform sediments: *Computers & Geosci.,* v. 15, no. 1, p. 95–105.

Flemings, P.B. and Jordan, T.E., 1987, Synthetic stratigraphy of foreland basins: *Eos Trans. Am. Geophysical Union,* v. 68, no. 16, p. 419.

Flemings, P.B. and Jordan, T.E., 1989, A synthetic stratigraphic model of foreland basin development: *Jour. Geophysical Research,* v. 94, no. B4, p. 3851–3866.

Haq, B.U., Hardenbol, J. and Vail, P.R., 1987, Chronology of fluctuating sea levels since the Triassic: *Science,* v. 235, no. 4793, p. 1156–1167.

Haq, B.U., Hardenbol, J., and Vail, P.R., 1988, Mesozoic and Cenozoic chronostratigraphy and cycles of sea-level change, *in* Wilgus, C.K., *et al.* (eds.), *Sea-Level Changes: An Integrated Approach:* Soc. Econ. Paleontologists and Mineralogists Spec. Pub. no. 42, p. 71–108.

Jervey, M.T., 1988, Quantitative geological modeling of siliciclastic rock sequences and their seismic expression, *in* Wilgus, C.K., *et al.* (eds.), *Sea-Level Changes: An Integrated Approach:* Soc. Econ. Paleontologists and Mineralogists Spec. Pub. no. 42, p. 47–69.

Kenter, J.A.M., 1991, Mineralogic composition and sediment fabric as controls on the geometry and declivity of submarine slopes, *in* Bosellini, A., *et al.* (eds.), *Dolomieu Conference on Carbonate Platforms and Dolomitization:* Ortisei/St.Ulrich, Val Gardena/Grödental, Dolomites, Italy, September 16–21, 1991, *Abstracts,* p. 137–138.

Kenyon, P.M. and Turcotte, D.L., 1985, Morphology of a delta prograding by bulk sediment transport: *Geol. Soc. Amer. Bull.,* v. 96, no. 11, p. 1457–1465.

Lawrence, D.T., Doyle, M. and Aigner, T., 1990, Stratigraphic simulation of sedimentary basins: Concepts and calibration: *Bulletin Am. Assoc. Petroleum Geologists,* v. 74, no. 3, p. 273–295.

Lerche, I., Dromgoole, E., Kendall, C.G.St.C., Walter, L.M. and Scaturo, D., 1987, Geometry of carbonate bodies: A quantitative investigation of factors influencing their evolution: *Carbonates and Evaporites,* v. 2, no. 1, p. 15–42.

McKenzie, D., 1978, Some remarks on the development of sedimentary basins: *Earth and Planetary Sci. Letters,* v. 40, no. 1, p. 25–32.

Parsons, B. and Sclater, J.G., 1977, An analysis of the variation of ocean floor bathymetry and heat flow with age: *Jour. Geophysical Research,* v. 82, no. 5, p. 803–827.

Scaturo, D.M., Strobel, J.S., Kendall, C.G.St.C., Wendte, J.C., Biswas, G., Bezdek, J. and Cannon, R., 1989, Judy Creek: A case study for a two-dimensional sediment deposition simulation, *in* Crevello, P.D., Wilson, J.L., Sarg, J.F. and Read, J.F., (eds.), *Controls on Carbonate Platform and Basin Development:* Soc. Econ. Paleontologists and Mineralogists Spec. Pub. no. 44, p. 63–76.

Schlager, W., 1981, The paradox of drowned reefs and carbonate platforms: *Geol. Soc. Amer. Bull.,* Part I, v. 92, no. 4, p. 197–211.

Sclater, J.G. and Christie, P.A.F., 1980, Continental stretching: An explanation of the post-mid-Cretaceous subsidence of the central North Sea basin: *Jour. Geophysical Research,* v. 85, no. B7, p. 3711–3739.

Sloss, L.L., 1982, Subsidence of continental margins: The case for alternatives to thermal contraction, *in* Scrutton, R.A., (ed.), *Dynamics of Passive Margins:* Am. Geophysical Union, Geodynamics Series, v. 6, p. 197–200.

Steckler, M.S. and Watts, A.B., 1978, Subsidence of the Atlantic-type continental margin off New York: *Earth and Planetary Sci. Letters,* v. 41, no. 1, p. 1–13.

Tetzlaff, D.M., 1986, Computer simulation model of clastic sedimentary processes: *Bulletin Am. Assoc. Petroleum Geologists*, v. 70, no. 5, p. 655.

Tetzlaff, D.M. and Harbaugh, J.W., 1989, *Simulating Clastic Sedimentation*: Van Nostrand Reinhold, New York, 202 pp.

Turcotte, D.L. and Kenyon, P.M., 1984, Synthetic passive margin stratigraphy: *Bulletin Am. Assoc. Petroleum Geologists*, v. 68, no. 6, p. 768–775.

Wildi, W., Funk, H., Loup, B., Amato, E. and Huggenberger, P., 1989, Mesozoic subsidence history of the European marginal shelves of the alpine Tethys (Helvetic realm, Swiss Plateau and Jura): *Eclogae geologicae Helvetiae*, v. 82, no. 3, p. 817–840.

Wilson, J.L., 1975, *Carbonate Facies in Geologic History*: Springer–Verlag, New York, 471 pp.

15

PRINCIPAL COMPONENT ANALYSIS OF THREE-WAY DATA

Michael Edward Hohn

Extension of conventional eigenvector analysis to three-way sets of data is possible through three-mode principal component analysis. First introduced in the 1960's, this method gives three sets of loadings corresponding to the three ways in the data, *e.g.*, variable, location, and time. A core matrix relates loadings across modes. Data must be centered and scaled before analysis, and as in conventional two-way analysis, preprocessing options affect the reduction in dimensionality and the appearance of the results. An example using water quality data illustrates the method and preprocessing effects. Although three-way tables can be studied through conventional analysis of a two-way table created by combining two modes of the data, three-mode analysis treats each mode separately and with the same weight. In addition, a restricted three-mode principal component model avoids problems in rotational indeterminacy, and results in a particularly simple model.

Applying Principal Component Analysis

Factor analysis or principal component analysis begins with a two-way table with samples along one margin and variables along the other. For instance, samples may be arranged as rows and variables as columns. R-mode analysis of the columns displays interdependencies among variables and Q-mode analysis displays similarities among samples. The analysis increases in complexity if a set of variables is repeatedly observed on the same samples; each set of measurements might represent a different experimental condition, chemical or sedimentological fraction, or simply geologic time. The resulting table of data can be visualized as a three-dimensional block: horizontal slices represent samples, vertical slices parallel with the front represent variables, and vertical slices parallel with the ends represent different conditions, times, or fractions. For example, Oudin (1970) performed elemental analyses of organic extracts from Jurassic shales in the Paris basin. Samples represented very different depths of maximum burial. Each was fractionated into several extracts according to solubility in organic solvents. Data published by Oudin (1970) have the three ways: locality, fraction, and element. Multivariate analysis of these data was presented by Hohn (1979).

Hohn and Friberg (1979) applied principal component analysis to petrographic data in which the three modes were sample, mineral phase, and chemical component. In gas chromatographic-mass spectral analysis of complex organic geochemical mixtures, sample, elution time (mass spectral scan number), and nominal mass formed three modes. Multivariate statistical analysis of crude oils, oil sands, and polluted beach sand created what appeared to be natural groupings of the oils on the basis of steranes and triterpanes (Hohn et al., 1981). Specifically, beach sands clustered with the type of crude spilled and laboratory-biodegraded Nigerian crude oils clustered with untreated samples of the same oils.

Three-way data recur in the geological literature. In a study of variation in shape of a Cretaceous foraminifer, Malmgren (1976) computed correlation matrices among growth-related variables from individuals within several stratigraphic intervals. Philp, Brown and Calvin (1978) measured organic content of photosynthetic bacteria and blue-green algae under increasingly severe thermal treatments. Lium et al. (1979) measured several chemical and biological variables at roughly one-month intervals over a year at four stations on the Chattahoochee River, Georgia. Methods for analyzing such data are not well known; statistical analyses have been limited to bivariate scattergrams and conventional eigenvector analysis. However, the method of three-mode principal component analysis provides a means for an integrated analysis of three-way data.

Three-mode principal component analysis was originally proposed in the psychometric literature (Tucker, 1966) and appeared useful in analysis of geological data (Hohn, 1979). Recent years have brought a number of applications in the natural sciences, notably in apportioning airborne contaminants to sources (Zeng and Hopke, 1990). In addition, several books and symposia treating the subject of multiway analysis appeared in the intervening years (Kroonenberg, 1983; Law et al., 1984). Papers by Kruskal (1984) and Harshman and Lundy (1984) describe in depth the options and effects of data scaling and centering on the eigenvector analysis. The purpose of this paper is three-fold: 1) to describe scaling and centering options more thoroughly than in earlier papers (Hohn, 1979; Hohn et al., 1981); 2) to provide an example that represents a future typical application of the method; and 3) to justify use of three-mode analysis and the collection of three-way data.

Methodology

Tucker (1966) described a method of analysis that he called "three-mode factor analysis." In the usual two-way table, a datum has one index corresponding to the sample and a second index corresponding to the variate; for example, x_{ij} refers to sample i and variable j. In three-mode analysis, x_{ijk} refers to the ith member of one mode, the jth member of a second mode, and the kth member of a third mode. A three-way principal component model can be written as

$$\tilde{x}_{ijk} = \sum_{p=1}^{P} \sum_{q=1}^{Q} \sum_{m=1}^{M} a_{ip} \, b_{jq} \, c_{km} \, g_{pqm} \, , \tag{1}$$

where $\tilde{\mathbf{X}}$ is a principal component approximation to the input data matrix $\mathbf{X}$; $\mathbf{A}$, $\mathbf{B}$, and $\mathbf{C}$ are three sets of eigenvectors; and $\mathbf{G}$ is a three-way "core matrix" as described below. If the data are pictured as a block composed of horizontal slices, they can be converted to a two-way format by arranging the slices side-by-side in a plane; when rows comprise samples, the first group of columns could contain measures taken for all variables on one occasion, the second group of columns would contain measures for all variables on a second occasion, *etc.* Together the columns are a combination mode having variables as an inner loop and occasions as an outer loop. Tucker (1966) represents the combination mode of i and j as (ij).

In the three-way model, matrix $\tilde{\mathbf{X}}$ will equal $\mathbf{X}$ exactly if the eigenvector matrices have a number of principal components equal to the number of entities in each mode, *i.e.*, $M = I$, $P = J$, and $Q = K$. However, in general, $M < I$, $P < J$, and $Q < K$ and $\tilde{\mathbf{X}}$ approximates $\mathbf{X}$.

One way to solve for the three-way model in Eq. (1) is to compute eigenvectors from an appropriate matrix of sums of squares and cross products. Eigenvectors A are calculated from a matrix $\mathbf{M}$ with elements

$$m_{uv} = \sum_{j=1}^{J} \sum_{k=1}^{K} x_{ujk} \, x_{vjk} \quad \text{for } u, v = 1, I \, . \tag{2}$$

Similarly,

$$r_{uv} = \sum_{i=1}^{I} \sum_{k=1}^{K} x_{iuk} \, x_{ivk} \quad \text{for } u, v = 1, J \tag{3}$$

and

$$s_{uv} = \sum_{i=1}^{I} \sum_{j=1}^{J} x_{iju} \, x_{ijv} \quad \text{for } u, v = 1, K \, , \tag{4}$$

from which sets of eigenvectors B and C are calculated.

Using graphical plots or other criteria, the analyst selects subsets of eigenvectors for each mode, so that $\mathbf{A}$ is $I \times M$, $\mathbf{B}$ is $J \times P$, and $\mathbf{C}$ is $K \times Q$. The $M \times P \times Q$ core matrix $\mathbf{G}$ is calculated from

$$g_{uvw} = \sum_{i=1}^{I} \sum_{j=1}^{J} \sum_{k=1}^{K} a_{iu} \, b_{jv} \, c_{kw} \, x_{ijk} \, . \tag{5}$$

Principal components for each mode can be rotated by the usual procedures, such as varimax (Kaiser, 1958). Rotation should precede calculation of the core matrix.

Eq. (1) shows that this model approximates an entry in $\mathbf{X}$ with the sum of $M \times P \times Q$ terms; each term is the result of multiplying appropriate loadings from each of the eigenvector matrices and weighting the product by the corresponding value in the core matrix. Levin (1965) and Kroonenberg (1984) call each eigenvector an "idealized" sample, variable, or condition. A sample that loads heavily on an eigenvector in the sample mode could act as an endmember in a gradient of samples; if it and several other samples load heavily on an eigenvector, they might constitute a cluster of very similar samples in multivariate space. In either circumstance, the eigenvector can be considered to represent an ideal sample. Gradients or clusters can occur among similar projections in the other two modes so that the total population of samples, variables, and conditions may be represented as a few, ideal entities. A given sample, variable, or condition might lie in gradients or clusters corresponding to several ideal samples, variables, or conditions, with the result that an observed value x_{ijk} can be regarded as a sum of all possible combinations of ideal entities among the modes. Variable j might give a high value for sample i under condition k; if entities i, j, and k load heavily on the axes m, p, and q of the respective modes, the result is high values for g_{mpq} in the core matrix. The pattern of scores in the core matrix reflects interdependencies across modes.

Scores

The usual model for principal component analysis may be written as

$$\tilde{x}_{ij} = \sum_{r=1}^{R} a_{ir} \, \lambda_{rr} \, b_{jr} \, , \tag{6}$$

where $\mathbf{X}$ is a two-way data array, $\mathbf{A}$ is an orthonormal eigenvector matrix of $\mathbf{X}\mathbf{X}'$, $\mathbf{B}$ is an orthonormal eigenvector matrix of $\mathbf{X}'\mathbf{X}$, and $_i\lambda_k$ is a diagonal matrix of square roots of eigenvalues with rows or columns appended if $i \neq k$. If mode i and j are variables and samples, the eigenvectors of the sample mode can be scaled by the eigenvalues,

$$\tilde{x}_{ij} = \sum_{r=1}^{R} a_{ir} \, s_{jr} \, , \tag{7}$$

where **A** comprises loadings of variables on i principal components and **S** comprises scores of samples on principal components.

The three-mode principal component model may be written as:

$$\tilde{x}_{ijk} = \sum_{p=1}^{P} a_{ip} \sum_{q=1}^{Q} \sum_{m=1}^{M} b_{jq} \, c_{km} \, g_{pqm} \, . \tag{8}$$

If the second and third modes represented by subscripts j and k are combined and treated as one mode subscripted with n, then:

$$\tilde{x}_{in} = \sum_{p=1}^{P} a_{ip} \, d_{np} \, , \tag{9}$$

where

$$d_{(jk)p} = d_{np} = \sum_{q=1}^{Q} \sum_{m=1}^{M} b_{jq} \, c_{km} \, g_{pqm} \, . \tag{10}$$

Thus, we see that matrix **D** is equivalent to **S**, and we can refer to matrix **D** as the scores of entities in the combination mode (jk) on P principal components. We can calculate the eigenvector matrices **A**, **B**, and **C**, then calculate corresponding score matrices. Results consist of six sets of coordinates; each pair of loading and score matrices makes up a conventional principal component analysis. The six matrices contain overlapping types of information.

Preprocessing

The data matrix **X** usually is centered and scaled before analysis. Harshman and Lundy (1984) refer to centering and scaling methods as *preprocessing*. Two-mode PCA or factor analysis usually involves centering the column variables so they have zero mean, and in factor analysis, scaling to unit variance by multiplying each column by the reciprocal of the column standard deviation. *Standardizing* is a combined centering to zero mean and scaling to unit variance. Preprocessing options include standardizing, scaling to unit mean-square (no centering), and scaling data between zero and one (Gower, 1966).

The options for preprocessing two-mode data are limited, and often dictated by the choice of R- or Q-mode analysis. In three-mode analysis, the researcher has the choice of centering and scaling from one to three modes, and some modes may be centered while others are scaled. Harshman and Lundy (1984) list reasons for preprocessing, including: 1) placing an emphasis in the analysis on change between entities within a mode by eliminating baseline values; 2) emphasizing particular subsets of the data over other subsets; 3) equalizing the influence of entities within a mode on the final solution; and 4) equalizing the influence of subsets of data on the final solution.

Baselines are eliminated by subtracting a minimum value (Hohn, *et al.*, 1981). Standardizing variables satisfies the third goal. The preprocessing method must be chosen based on the objective of the analysis. Preprocessing affects the dimensionality of the solution, and thus interpretability of results. If preferable methods reduce the number of components or factors in the final model, we can use the findings of Kruskal (1984) and Harshman and Lundy (1984) to select among preprocessing options. Particular preprocessing strategies are more effective in reducing the size of the solution than others:

1. Centering one-way subarrays of data reduces dimensions. More reduction is achieved by double-centering, or centering subarrays across one mode:

$$x_{ijk}^* = x_{ijk} - \frac{1}{I} \sum_{i=1}^{I} x_{ijk} \qquad (11)$$

and then centering across a second mode:

$$x_{ijk}^\circ = x_{ijk}^* - \frac{1}{J} \sum_{i=1}^{J} x_{ijk}^* . \qquad (12)$$

The order of the centering is not important.

2. Two-way centering involves two-way subarrays of the data:

$$x_{ijk}^* = x_{ijk} - \frac{1}{IJ} \sum_{i=1}^{I} \sum_{j=1}^{J} x_{ijk} \qquad (13)$$

and does not yield a reduction in dimensionality. Two-way centering is not equivalent to double centering.

3. Rescaling one-way subarrays does not reduce dimensionality.

4. Rescaling two-way subarrays reduces the number of dimensions. More than one mode may be rescaled by an iterative process of rescaling one mode, then the second, then the first again, continuing until convergence, usually three to six cycles.

Example: Water Quality

Lium *et al.* (1979) measured algal growth potential (AGP) (mg/L), phytoplankton concentration (cells/ml), average lake temperature and dissolved orthophosphate, and nitrate concentrations (mg/L) over a period of almost

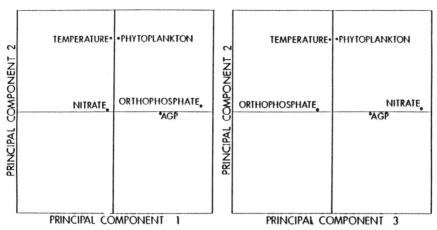

Figure 1: Results of three-mode principal component analysis of water quality data; variable mode.

a year at four stations in West Point Lake, an artificial Lake on the Chattahoochee River in Georgia, USA. The purpose of the study was to evaluate use of biological and microbiological data in estimating water quality, to assess potential for algal growth, and to determine controls on biological and microbiological communities. Before three-mode principal component analysis, values within each location–variable combination were centered and the sums of squares and cross products scaled to unity (Table 1). The removed additive differences among locations equalized the influence of each variable.

Water Quality Loadings

Unrotated principal components and eigenvalues (Table 2) showed that two or three axes adequately explained variability in the data. Three axes were rotated for the variable and time modes, and two axes for the location mode. Average lake temperature and phytoplankton concentration are highly correlated (Fig. 1), as noted by Lium *et al.* (1979). These variables are nearly orthogonal to the plane containing orthophosphate, nitrate, and AGP. AGP decreases and concentration of phytoplankton increases downstream except when AGP falls below 0.5 mg/L; because the relationship between AGP and phytoplankton is curvilinear, correlation between these two variables is low. In addition, the centering of phytoplankton counts within each site may have obscured a geographic trend. Lium *et al.* (1979) were able to predict AGP from nitrate and orthophosphate concentrations. Figure 1 indicates that no correlation exists between nitrate and orthophosphate, and that AGP is related to both of these nutrients.

Ordination of locations on two principal axes (Fig. 2) shows a gradient from the most upstream site, Franklin, to the most downstream site, the dam

Table 1: Water Quality Data from West Point Lake, Georgia,
(1) After Centering and Scaling Each Location–Variable Combination;
(2) After Centering and Scaling Each Variable

Location (river mi) and Collection Date	Aver. Lake Tempera- ture (°C) (1)	(2)	Nitrate (mg/L) (1)	(2)	Ortho- phosphate (mg/L) (1)	(2)	Algal Growth Potential (mg/L) (1)	(2)	Phyto- plankton (log cells/ml) (1)	(2)
Franklin (0.0)										
1. Dec. 1975	−.47	−.23	.11	.23	.04	.27	.17	.30	−.17	−.20
2. Feb. 1976	−.40	−.20	−.34	.09	.04	.27	−.16	.19	−.14	−.20
3. March	−.41	−.21	−.59	.01	−.73	−.06	−.69	.01	−.43	−.28
4. April	−.03	−.02	−.27	.11	−.34	.11	−.20	.18	−.37	−.27
5. July	.28	.14	.00	.20	.23	.36	−.04	.23	.35	−.04
6. Aug. 10	.37	.19	.28	.29	.32	.40	.30	.34	.67	−.06
7. Aug. 24	.42	.21	.28	.29	.04	.27	.05	.26	.21	−.09
8. Sept.	.24	.12	.54	.37	.42	.44	.58	.44	−.12	−.19
LaGrange (14.2)										
1. Dec. 1975	−.47	−.23	−.16	.01	−.20	−.06	−.43	−.08	−.26	−.14
2. Feb. 1976	−.40	−.20	.04	.07	.49	.07	.11	.06	−.28	−.15
3. March	−.41	−.20	−.40	−.07	−.43	−.10	−.42	−.08	−.59	−.33
4. April	−.03	−.02	−.58	−.12	−.43	−.10	−.47	−.09	−.16	−.08
5. July	.28	.14	.13	.10	.26	.03	.32	.12	.33	.20
6. Aug. 10	.37	.19	.45	.19	.48	.07	.42	.15	.48	.28
7. Aug. 24	.42	.21	.00	.05	.03	−.01	.27	.10	.37	.22
8. Sept.	.24	.12	.51	.21	−.20	−.06	.21	.09	.11	.07
Abbottsford (24.8)										
1. Dec. 1975	−.46	−.23	.56	.08	.13	−.10	.00	−.12	−.28	−.06
2. Feb. 1976	−.40	−.20	.50	.06	.85	−.01	.86	−.04	−.33	−.09
3. March	−.41	−.20	−.07	−.12	.13	−.10	.15	−.11	−.61	−.24
4. April	−.03	.02	−.30	−.19	−.22	−.14	−.05	−.13	.00	.09
5. July	.29	.14	−.57	−.29	−.22	−.14	−.13	−.14	.28	.23
6. Aug. 10	.37	.19	−.05	−.12	−.22	−.14	−.29	−.16	.27	.23
7. Aug. 24	.42	.21	−.03	−.11	−.22	−.14	−.25	−.15	.51	.35
8. Sept.	.24	.12	−.05	−.12	−.22	−.14	−.29	−.16	.15	.17
Dam Pool (33.1)										
1. Dec. 1975	−.47	−.23	.54	.03	−.20	−.14	.19	−.14	−.43	−.03
2. Feb. 1976	−.40	−.20	.41	−.01	.61	−.10	.85	−.12	−.30	−.00
3. March	−.41	−.20	.31	−.05	−.20	−.14	−.05	−.15	−.52	−.04
4. April	−.03	−.02	.04	−.14	−.20	−.14	−.05	−.15	.03	.06
5. July	.28	.14	−.41	−.29	−.20	−.14	−.14	−.15	.19	.10
6. Aug. 10	.37	.19	−.39	−.28	.61	−.10	−.28	−.16	.15	.09
7. Aug. 24	.42	.21	−.28	−.24	−.20	−.14	−.24	−.16	.52	.16
8. Sept.	.24	.12	−.22	−.22	−.20	−.14	−.28	−.16	.36	.13

Table 2: Unrotated Principal Components and Core Matrix from Water Quality Data after Centering and Scaling Each Location–Variable Combination

Mode	Principal Component	Eigen-value	Cum. Percent	Core Matrix				
				Loc.	Time	Variable		
						1	2	3
Variable	1	9.63	48	1	1	1.61	−1.20	1.03
	2	7.60	86		2	−.32	−.91	.11
	3	1.75	95		3	.16	−1.19	1.05
Location	1	11.97	60	2	1	−.01	−.87	−.71
	2	6.15	91		2	−1.52	−1.11	−1.33
Time	1	12.71	64		3	−.39	−1.26	−.09
	2	4.09	84					
	3	1.72	93					

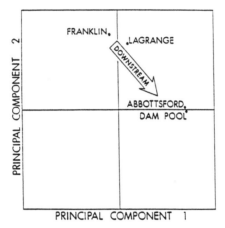

Figure 2: Results of three-mode principal component analysis of water quality data; location mode.

pool. Time does not exhibit the obvious gradient or cycle expected (Fig. 3), although the third axis can be interpreted as a gradient from winter and spring months (positive loadings) to summer months (negative loadings). With the second axis, these loadings describe a cyclic pattern with irregularities at the second and seventh months. February appears anomalous in having a high loading on the first axis.

Water Quality Core Matrix

Rows in the core matrix (Table 2) corresponding to the first time component contain high positive or negative values for all three variable components on the first location component. The highest loading in the core matrix (1.61) reflects the high AGP, orthophosphate and nitrate concentration, and low

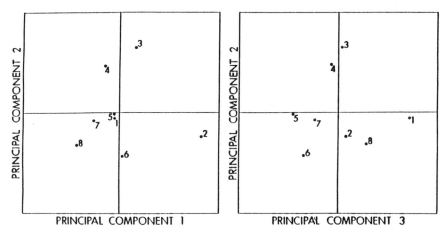

Figure 3: Results of three-mode principal component analysis of water quality data; time mode.

temperature and phytoplankton counts at Abbottsford and the dam pool during February.

The core matrix can be used to explain salient features in one of the graphs of principal components; conversely, salient features or distinctive patterns of entries in the core matrix can be interpreted from the loadings on principal components. For example, the third variable component, representing nitrate and AGP, has positive weight for the third time component, first location component and negative weight for the second time component, second location component. The positive value reflects high nitrate concentration and AGP in December at Abbottsford and the dam pool. The negative value corresponds to low values for these variables in March and April at Franklin and LaGrange. The three negative loadings in the fifth row of the core matrix indicate that all variables reach low values in spring at upstream sites. The block of negative loadings in the second column of the core matrix indicates there are no major contrasts for phytoplankton count among locations on a given collection date. Principal components for collection dates are bipolar, so weights in this row of the core matrix have equal magnitudes and signs, but contrasts do exist between collection dates in phytoplankton count; centering preserved such contrasts among dates, but may have eliminated mean contrasts among locations.

Scores for Combination of Location and Time

A matrix of scores was calculated for the 32 combinations of location and time on two principal components. The plot of the scores shows distinct effects attributable to geographic location and time of year. In Figure 1 the first principal component in variable mode conforms to nitrate–ortho-

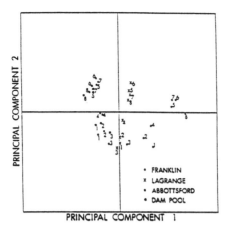

Figure 4: Scores of location–collection date combinations on principal components.

phosphate–AGP and the second component conforms to phytoplankton–temperature. The major effect on phytoplankton counts appears to be time of year (Fig. 4), although geographic effects can be discerned in summer months. The nutrient component responds to a geographic effect—expressed by the mean score of points from each location on the first axis—and a time-of-year interaction with the collection site. Note the change in orientation of points representing sites; summer months score more positively on the first component than winter and spring months at the Franklin site, whereas the contrast is reversed at downstream sites. This is observed in the core matrix.

Effects of Different Scaling

Lium *et al.* (1979) noted a decrease in phytoplankton counts downstream from Franklin; centering and scaling within the variable mode (Table 1) should preserve this geographic trend. This constitutes two-way centering and scaling because each location–time subarray is adjusted so the sum of squares and cross products among variables forms a correlation matrix. Two principal components from each mode (Table 3; Fig. 5) are adequate for reconstructing the data and resemble previous results except for higher correlations among AGP, nitrate, and orthophosphate and a well-defined gradient among collection times.

The core matrix (Table 3) shows a simple pattern of signs and magnitudes. The first variable component has positive weights for the first location component: AGP and related variables peak in the upstream locality, especially in summer months. These variables decrease downstream and show lower values in summer relative to winter and spring. The second column

Table 3: Unrotated Principal Components and Core Matrix from Water Quality Data after Centering and Scaling Each Variable

Mode	Principal Component	Eigen-value	Cum. Percent	Loc.	Time	Variable 1	2
Variable	1	2.94	59	1	1	1.19	.40
	2	1.65	92		2	.50	−.73
Location	1	2.85	57	2	1	−.81	.86
	2	1.88	95		2	−.57	−.51
Time	1	3.16	63				
	2	1.41	92				

corresponds with a temperature–phytoplankton component, showing high values in summer relative to winter and spring.

Why Three-Mode Analysis?

Why use such an apparently complex method as three-mode analysis when simpler methods are available? The three-way raw data could be studied as a series of conventional two-mode analyses or two modes might be combined into one. However, three-mode analysis may be preferable when none of the modes is the logical one for breaking the table up into two-way tables or when the analysis requires study of principal component loadings for all three modes. If it is not necessary to relate principal components across modes, the complexity of interpreting the core matrix can be avoided. This is usually not the case. However, scores help interpret the core matrix. For example, in gas chromatographic-mass spectrometry, a complex organic mixture in solution is separated in a gas chromatograph and continuously fed as a gas into a mass spectrometer, which analyzes for the quantity of material having a specified mass-to-charge ratio. The raw data has chromatographic retention time (expressed as scan number) as one mode, nominal mass as a second mode, and samples as the third mode. Hohn *et al.* (1981) noted the following advantages in using principal component analysis and three-mode PCA:

1. Principal components of nominal mass represent linear combinations of the raw data. This partially mitigates the effects of error in nominal mass value.

2. The analyst can discern clusters of samples and determine which nominal masses account for contrasts and similarities among clusters.

3. Multivariate analysis precludes the need to fully resolve and quantify compounds in these complex mixtures. Principal component loadings

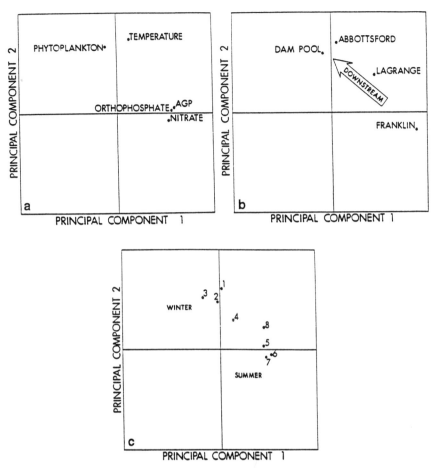

Figure 5: Three-mode principal component analysis of water quality data, scaled by variables: (a) variable mode, (b) location mode, (c) time mode.

of mass spectral scans plotted against scan number can be used as fingerprints for classifying samples.

4. In the study by Hohn *et al.* (1981), it was known which nominal masses would probably result in a logical grouping of samples. An expected group of masses characteristic of triterpanes and a second group characteristic of steranes were obseved; steranes and triterpanes were the major components of the analyzed organic fractions. The plots confirmed that samples grouped along expected criteria.

In summary, three-mode analysis is appropriate for three-way data when the researcher wants to examine components in each mode, and wants to relate components across modes. The richness of information obtained from a three-mode analysis suggests that the extra time and work necessary to collect multi-way data may be justified by the results.

References

Gower, J.C, 1975, Generalized Procrustes analysis: *Psychometrika*, v. 40, no. 1, p. 35–51.

Harshman, R.A. and Lundy, M.E., 1984, Data preprocessing and the extended PARAFAC model, *in* Law, H.G., Snyder, C.W., Hattie, J.A. and McDonald, R.P., (eds.), *Research Methods for Multimode Data Analysis:* Praeger, New York, p. 216–284.

Hohn, M.E, 1979, Principal components analysis of three-way tables: *Math. Geol.*, v. 11, no. 6, p. 611–626.

Hohn, M.E. and Friberg, L.M., 1979, A generalized principal components model in petrology: *Lithos*, v. 12, p. 317–324.

Hohn, M.E., Jones, N.W. and Patience, R., 1981, Objective comparison of mass fragmentograms by principal components analysis: Method and results: *Geochim. et Cosmochim. Acta*, v. 45, p. 1131–1140.

Kaiser, H.F., 1958, The varimax criterion for analytic rotation in factor analysis: *Psychometrika*, v. 23, no. 3, p. 187–200.

Kroonenberg, P.M., 1984, Three-mode principal component analysis illustrated with an example from attachment theory, *in* Law, H.G., Snyder, C.W., Hattie, J.A. and McDonald, R.P., (eds.), *Research Methods for Multimode Data Analysis:* Praeger, New York, p. 64–103.

Kroonenberg, P.M., 1983, *Three-Mode Principal Component Analysis: Theory and Applications:* DSWO Press, Leiden, 398 pp.

Kruskal, J.B., 1984, Multilinear methods, *in* Law, H.G., Snyder, C.W., Hattie, J.A. and McDonald, R.P., (eds.), *Research Methods for Multimode Data Analysis:* Praeger, New York, p. 36–62.

Law, H.G., Snyder, C.W., Hattie, J.A. and McDonald, R.P., (eds.), 1984, *Research Methods for Multimode Data Analysis:* Praeger, New York, 686 pp.

Levin, K., 1965, Three-mode factor analysis: *Psychol. Bull.*, v. 64, no. 6, p. 442–452.

Lium, B.W., Stamer, J.K., Ehlke, T.A., Faye, R.E. and Cherry, R.N., 1979, Biological and microbiological assessment of the upper Chattahoochee River Basin, Georgia: *Prof. Circ. 796*, U.S. Geological Survey, 22 pp.

Malmgren, B.A., 1976, Size and shape variation in the planktonic foraminifer *Heterohelix striata* (Late Cretaceous, Southern California): *Math. Geol.*, v. 8, no. 2, p. 165–182.

Oudin, J.L., 1970, Analyses Géochimique de la matière organique extraite des roches sédimentaires. I. Composés extractibles au chloroforme: *Inst. Français du Pétrole Rev.*, v. 25, no. 1, p. 3–15.

Philp, R.P., Brown, S. and Calvin, M., 1978, Isoprenoid hydrocarbons produced by thermal alteration of *Nostroc moscorum* and *Rhodopseudomonas spheroides*: *Geochim. et Cosmochim. Acta*, v. 42, no. 1, p. 63–68.

Tucker, L.R., 1966, Some mathematical notes on three-mode factor analysis: *Psychometrika*, v. 31, no. 3, p. 279–311.

Zeng, Y. and Hopke, P.K., 1990, Methodological study applying three-mode factor analysis to three-way chemical data sets: *Chemometrics and Intelligent Laboratory Systems*, v. 7, p. 237–250.

16

A SOLUTION TO THE PERCENTAGE-DATA PROBLEM IN PETROLOGY

E. H. Timothy Whitten

Statisticians have demonstrated the inappropriateness of percentage data for petrological purposes except when transformations (*e.g.*, log-ratios) are used to avoid inherent closure. Use of open variables for chemical data (perhaps weight per unit volume, g/100cc) would avoid this problem and permit traditional petrological work to be undertaken.

The Closure Problem

Virtually all compositional data used in petrology are expressed as percentages (*e.g.*, SiO_2 wt%, muscovite vol%, Si wt%) or parts per million. Geologists depend on percentage and ppm data for studies of petrogenesis, spatial variability, *etc.* However, for over four decades, statisticians and mathematical geologists have given dire warnings about the dangers of drawing conclusions from percentage (or ratio) data. In consequence, petrological literature abounds with disclaimers about possible adverse effects that closure constraints (stemming from use of percentage data) *may* have.

The abundant warnings have given little help to geologists for two reasons. First, the precise impact of closure on petrologic analyses and conclusions has been unclear or abstract. Second, a practical and realistic way of avoiding closure in petrology has not been apparent. Problem avoidance might involve either (a) applying statistical or mathematical transformations to standard percentage data to escape the inherent closure constraints that importune petrological conclusions, or (b) using meaningful petrological variables that are free of closure constraints so that traditional thinking and data manipulation can be used without problem.

Transformation has been advocated for geological work by Aitchison (*e.g.*, 1982); this approach, which presents considerable geological difficulties, is briefly reviewed here. No attention appears to have been given to the simple approach of using closure-free variables, which is the main subject of this paper.

Closed (Percentage) Data and Petrology

Closed data are compositional data that have a constant sum. Open data can have any value and do not have the constant-sum constraint. Standard rock chemical analyses are closed because the oxides (or elements, *etc.*), expressed as percentages, sum to 100; in consequence, at least one negative correlation between the variables must exist (Chayes, 1960). The problem of closure is obvious in two-variable systems. In a quartz-feldspar rock, for example, if quartz percentage increases, feldspar must decrease, so there is inherent negative correlation between the components. With more than two components inherent correlation relationships persist but become more complex as the number of variables increases (Chayes, 1971). Most standard (classical) statistical tests used in petrology assume normally distributed data and independence of the variables involved.

Chayes (1971), Aitchison (1982, 1986), and others demonstrated forcefully that it is erroneous statistically to use percentage chemical data for many standard petrographical tasks. Relationships and trends on standard petrological Harker and ternary diagrams are particularly vulnerable. Aitchison (1990, p. 487) correctly affirmed "... the accumulated evidence of the last 40 years, starting with Chayes (1948, 1960) and Sarmanov and Vistelius (1959), that 'standard' statistical concepts, such as product-moment correlation, developed for the analysis and interpretation of unconstrained data, are inapplicable to constrained data such as rock compositions subject to a constant-sum constraint."

While caution is now common when treating major rock constituents, many petrologists assume trace-element relationships, although part of the constant sum, are not impacted by closure. However, Chayes (1983) showed that if pre-closure variances of two trace elements are sufficiently small their closed equivalents will be positively correlated, so that some commonly observed positive correlations between trace elements may be products of closure (*cf.* Rock, 1988, p. 204).

Statistical studies have shown clearly what should not be done with closed geological data.

Transforming Percentage Data to Avoid Closure Problems

In a series of papers, Aitchison demonstrated how percentage compositional data can be transformed appropriately for statistical analysis. In dealing with percentage data, Aitchison (1986, p. 65; 1990, p. 487) noted that the study of rock compositions is essentially concerned with the relative magnitudes, or ratios, of ingredients, rather than their absolute values, and that it is not meaningful, therefore, to consider components in isolation. Hence, he asserted that one should think in terms of ratios and a concept of correlation structure based on product-moment covariances of ratios. He pointed out that, following standard statistical practice, this is simpler after transforming raw analyses to log-ratios—for example, each variable from a closed array may be divided by a particular variable (*e.g.*, divide other major-oxide percentages by silica or MgO percentage)—and taking logarithms of the resulting ratios which yields a set of unrestricted open data. Stanley (1990, fig. 1) developed a Harker-type graph that is helpful in putting these relationships into clear geological perspective.

Starting with numerous sets of published geological percentage data, Aitchison used log-ratios to escape closure, perform statistically appropriate analyses, and conclude that earlier petrographical interpretations need significant change. Woronow and Love (1990) adapted this approach to quantify and test the differences among means of chemical compositional data, and Zhou *et al.* (1991) adapted it to partition modern sediments (using grain size) in the northern South China Sea.

Manipulating transformed geological data yields results in transformed-data space; commonly, it is difficult to interpret such results in terms of the original data (*cf.* N. I. Fisher *in* Aitchison, 1982, p. 165). Because transformed log-ratio data and original open data comprise two different sets of variables, the same partitioning[1] ("classification") technique (*e.g.*, cluster analysis) applied to each set of variables will, in general, yield very dissimilar results. Different, mathematically correct partitionings result from use of different sets of variables (for the same samples), so the geologist must determine which variables are relevant to meaningful partitioning. While the use of closed data is generally inappropriate, use of log-ratio data may yield mathematically correct, but geologically irrelevant partitioning or other results.

If percentage data are used—as is traditional in petrology (and much of the rest of geology)—the transformation that Aitchison advocated [or some other suitable transformation such as, for example, addition of white noise (Sharp, 1989)] is mandatory. Transformation is the price to pay for using

---

[1]Strictly, most igneous-rock "classification" is really only partitioning (see Whitten, 1987, and Whitten *et al.*, 1987, for definitions of partitioning and classification).

such compositional data. Aitchison (1990) referred to studies of composition having to be concerned with relative values; this is true for the chemical variables used most commonly in petrology, but *not* for open compositional variables—possibly g/100cc, for example. The latter appears to provide a preferable approach that is free from closed-array constraints.

When equal-volume metasomatic changes have been evaluated, chemical composition of equal volumes has been assessed in g/100cc (*e.g.*, Wager, 1929; Exley, 1959; Boyle, 1961). Tyrrell (1948, 1952) used weight per unit volume to evaluate correctly the spatial compositional variability of the differentiated Lugar Sill. Whitten (1972, 1975) advocated that, for most purposes, oxide (or cation) compositions of igneous rocks be expressed in g/100cc, rather than wt%. Where spatial chemical variability within a rock unit is of interest, the absolute amount of component in similar volumes is of interest, rather than the wt% (*i.e.*, amounts in samples of dissimilar size but equal weight).

Available Data

Grams per 100 cc (g/100cc) is wt% multiplied by whole-rock specific gravity. Quality whole-rock chemical analyses are abundant in the literature, but few authors record specific gravity of analyzed samples. Whitten (1975, tables 2 and 4) listed some sources of major-element and specific-gravity data. The paucity of specific-gravity data probably reflects the perceived difficulty of measurement. However, at Pomona College, California, McIntyre *et al.* (1965) demonstrated the accuracy of the air-comparison pycnometer for whole-rock specific-gravity analysis, and, concomitantly, Baird *et al.* (*e.g.*, 1962, 1967a) developed accurate X-ray spectrographic methods for whole-rock chemical analysis. Although technical improvements in chemical analytical methods have since been developed elsewhere, very few recent chemical data are supported by specific-gravity data. The results recorded by A.K. Baird, D.B. McIntyre and their associates for southern California granitoids are undoubtedly the best combined specific gravity and chemical information available currently in the literature.

Here, 154 of the 162 analyses for Lakeview Mountains Pluton (Morton *et al.*, 1969) and three geographic subsets of the approximately 550 Northern Peninsula and Transverse Ranges batholiths analyses (Baird *et al.*, 1979) are used—eight or nine major element weight percentages plus whole-rock specific gravity were recorded for each sample. [Unfortunately, rather than being included in the printed paper, the 162 Lakeview Mountains Pluton analyses were stored on chemically treated paper in a depository and A.K. Baird is now deceased; while most of these data can be read easily, for eight samples (LV-11, -61, -104, -105, -109, -110, -116, -156), one arbitrary entry is now unreadable, so these eight samples have been omitted here.]

Table 1: Specific Gravity Variability in Available Samples

	San Gabriel Mountains	Little San Bernardino Mountains	Santa Ana Mountains	Lakeview Mountains
Samples	47	25	45	154
Mean value	2.735	2.707	2.735	2.817
Range	2.644–2.835	2.644–2.809	2.637–2.902	2.733–2.926
χ^2	5.54	11.23	31.81*	15.76
$\chi^2_{(9)}$	16.92	16.92	16.92	16.92

*Not normally distributed

Specific Gravity *vs.* Chemical Composition

Specific gravity of samples from a rock unit commonly shows considerable variability. Table 1 summarizes variability within samples used in this study; the χ^2-test suggests normal distributions except for the Santa Ana Mountains granitoids.

Because specific gravity is, in general, different for each plutonic rock sample, the sum of weights of all chemical components (in g/100cc) is different for each sample. That is, weights per unit volume appear to be open data unconstrained by the inherent statistical correlations associated with wt% data. This would mean that g/100cc is a variable of choice (rather than weight percentages) for depicting petrographic spatial variability and for most petrogenetic calculations and correlations.

The independence of the g/100cc variables (oxides, elements, *etc.*) might be compromised, however, if the specific gravity of individual rock samples were directly correlated with the wt% of the constituents. Intuitively, strong correlation seems unlikely in granitoids because of (OH) which, although not included in most chemical analyses, has significant effects on mineral specific gravity (micas, amphiboles, *etc.*).

For the 154 Lakeview Mountains Pluton samples used, linear correlation between specific gravity and individual elements/oxides is not strong (see Table 2); r values are identical when oxide weight percentages and oxide g/100cc values are used instead of element wt% and g/100cc, respectively. If linear-regression lines are used to predict rock specific gravity from each element (or oxide) weight percentage, Mg wt% (or MgO wt%) accounts for only 55.65% of the total sum of squares of the variability, but this is more than for any other element (or oxide). That is, whole-rock density is not a simple function of element or oxide composition.

Table 3 shows linear correlation coefficients between whole-rock specific gravity and elemental analyses for granitoids from the Northern Peninsula

Table 2: Lakeview Mountains Pluton ($N = 154$) – Linear
Correlation (r) between Each Element and
Whole-Rock Specific Gravity

Element	Using wt%	Using g/100cc
Si	−0.717	−0.486
Al	0.138	0.356
Fe	0.583	0.646
Mg	0.746	0.781
Ca	0.621	0.682
Na	0.153	0.348
K	−0.550	−0.524
Maximum sum of squares accounted for by one variable (Mg)	55.65%	61.00%

and Transverse Ranges. Linear-regression lines to predict whole-rock density from each element weight percentage for these data account for more of the total sum of squares than do those for Lakeview Mountains Pluton. However, the "best" predictor element is different for each sample set. For San Gabriel Mountains less than 80% of the total sum of squares is accounted for by any element (Table 3), so specific gravity cannot be calculated accurately from any element weight percentage. The Santa Ana Mountains data yield larger r values, but these are misleading because χ^2-tests indicate these data are not normally distributed (see Tables 1 and 4).

Another question with g/100cc data is whether, because 100 cc is a constant, individual oxides (or cations) are actually independent (*cf.* Whitten, 1975, p. 304). This is a complex issue which is examined here in relation to an arbitrary specimen, San Gabriel Mountains granitoid sample B1. Baird *et al.* (1979) reported the analysis in element wt% (Table 5, column 1). By allocating requisite oxygen, these data are readily converted to oxide wt% (column 2) and, as is common for published data, the results are re-percentaged in column 3 (compounding further the closure problem). Column 4 gives equivalent g/100cc values based on column 3, although use of column 2 yields more correct results (column 5).

Without actual chemical experiment, or very complex thermodynamic calculations, it is difficult to know precisely what would happen if some components in column 5 were now changed. However, possibilities can be examined briefly. If 1.5 g FeO and 0.5 g MgO are added to the assemblage, they could probably be accommodated in the 100 cc with specific gravity increasing from 2.696 to 2.716, and all other oxides remaining unchanged (column 6) until the new "rock" is expressed as weight percentages

Table 3: Linear Correlation, r, between Each Element (Using wt%
and g/100cc Data) and Whole-Rock Specific Gravity

Element	San Gabriel Mtns ($N = 47$) wt%	San Gabriel Mtns ($N = 47$) g/100cc	Little San Bernardino Mtns ($N = 25$) wt%	Little San Bernardino Mtns ($N = 25$) g/100cc	Santa Ana Mtns ($N = 80$; no gabbros) wt%	Santa Ana Mtns ($N = 80$; no gabbros) g/100cc
Si	−0.874	−0.789	−0.926	−0.871	−0.945	−0.906
Al	0.838	0.893	0.535	0.695	0.870	0.913
Fe	0.859	0.869	0.897	0.901	0.958	0.965
Mg	0.803	0.811	0.852	0.859	0.943	0.947
Ca	0.864	0.873	0.924	0.929	0.942	0.949
Na	0.277	0.492	0.105	0.263	−0.508	−0.305
K	−0.778	−0.749	−0.766	−0.734	−0.827	−0.816
Ti	0.820	0.834	0.903	0.909	0.919	0.929
Max SS*	Si 76.39	Al 79.74	Si 85.75	Ca 86.30	Fe 91.78	Fe 93.12

*Maximum sum-of-squares percentage accounted for by one
variable (the element being indicated in each case)

(column 7). Assuming the P,T domain remains unchanged, but allowing for possible variation of unrecorded (OH), CO_2, *etc.*, the specific gravity and volume of the mineral aggregate will change (from those shown in column 6) due to chemical re-equilibration of the several mineral phases.

If larger additions to, or subtractions from, the original composition are postulated, the initial volume must change. For example, if 20 g of SiO_2 were added to the already SiO_2-saturated sample, a volume increase to, perhaps, 107.55 cc (specific gravity of quartz being 2.65) is reasonable (column 8), which yields the weight percentages in column 9. In real rocks, the apparent constraint of constant volume is not real and the grams of oxide (element) can be changed (in g/100cc space) without impacting the others' abundance (until re-expressed as wt%). If 20 g of $CaCO_3$ were added to sample B1 (column 5), the initial 100 cc would probably increase, but, because C is not normally recorded (and/or might be lost as CO_2 from any

Table 4: χ^2 Values for Selected Variables

Variable	San Gabriel Mountains	Little San Bernardino Mtns	Santa Ana Mountains	Lakeview Mountains
Si	3.67	5.42	19.25*	8.98
Fe	11.76	14.33	22.28*	16.23
Mg	28.50*	34.84*	47.17*	13.09
Ca	5.31	7.87	34.32*	8.36
$\chi^2_{(9)}$	16.92	16.92	16.92	16.92

*Not normally distributed

Table 5: Granitoid Sample B1, San Gabriel Mountains, California

E*	1	2	3	4	5	6	7	8	9	10	11
Si	33.06	70.73	71.20	191.96	190.68	190.68	70.21	210.68	72.74	190.68	65.84
Al	8.10	15.31	15.41	41.55	41.26	41.26	15.19	41.26	14.25	41.26	14.25
Fe	1.93	2.48	2.50	6.74	6.69	8.19	3.02	6.69	2.31	6.69	2.31
Mg	0.52	0.86	0.87	2.35	2.32	2.82	1.04	2.32	0.80	2.32	0.80
Ca	1.62	2.27	2.28	6.15	6.11	6.11	2.25	6.11	2.11	17.31	5.98
Na	3.05	4.11	4.14	11.16	11.08	11.08	4.08	11.08	3.83	11.08	3.83
K	2.61	3.14	3.17	8.55	8.48	8.48	3.12	8.48	2.93	8.48	2.93
Ti	0.26	0.43	0.44	1.19	1.17	1.17	0.43	1.17	0.40	1.17	0.40
Ue	48.85	0.67	0.00	0.00	1.81	1.81	0.66	1.81	0.62	10.61	3.66
Σ	100.00	100.00	100.00	269.65	269.60	271.60	100.00	289.60	100.00	289.60	100.00
cc	na	na	na	100.00	100.00	100.00	na	107.55	na	107.38	na
sp	2.696	2.696	2.696	2.696	2.696	2.716	2.716	2.693	2.693	2.697	2.697

*Captions for Table 5. E – The eight elements or oxides (Ue = unanalyzed components + analytical errors);
Σ – Sum; cc – Volume in cc; sp – Specific gravity.

1 – Element wt% (from Baird et al., 1979).
2 – Oxide wt%.
3 – Re-percentaged oxide wt%.
4 – Oxide g/100cc (based on column 3).
5 – Oxide g/100cc (based on column 2).
6 – Addition of 1.5 g FeO and 0.5 g MgO to column 5.
7 – Oxide wt% from column 6.
8 – Addition of 20 g SiO_2 to column 5.
9 – Oxide wt% from column 8.
10 – Addition of 20 g $CaCO_3$ to column 5.
11 – Oxide wt% from column 10.

natural system), both total grams and sample volume would increase (CaO by 11.20 g and volume by 7.38 cc, say). Re-expressed as weight percentages, CaO would be increased and all other oxides decreased (column 11, which assumes CO_2 is retained in the system).

While not an exhaustive analysis, these calculations show that, when adding to (or subtracting from) components of a rock expressed in g/100cc units, specific gravity and/or volume of the aggregate vary (varies) due to recrystallization, effects of (OH), CO_2, etc., in a manner unpredictable from g/100cc data alone. This supports the contention that actual petrological g/100cc data are open (i.e., not constrained by closure).

Conclusions

For the southern California granitoid data tested, whole-rock specific gravity is not a simple function of chemical composition and weight-per-unit volume chemical data appear to be unconstrained open variables. It is reasonable to anticipate that g/100cc chemical values for major and trace elements in other igneous rocks show similar characteristics. Because g/100cc data are directly meaningful to petrologists, their apparent openness should be studied exhaustively.

Considerable detail has been recorded about the sampling, chemical variability, and spatial variability of Lakeview Mountains Pluton (e.g., Baird et al., 1967b; Morton et al., 1969); it would be useful to employ g/100cc data in statistical evaluation of this information.

Patterns in spatial-variability maps for g/100cc can be expected to be generally similar to those for weight percentages, as is the case for Al_2O_3 wt% and g/100cc trend-surface maps for part of Thorr Pluton, Eire (Whitten, 1975, fig. 11).

Distributions on Harker-type diagrams also tend to be broadly similar when wt% and g/100cc data are used. Table 6 shows correlation-coefficient matrices for San Gabriel Mountains granitoid wt% and g/100cc data. The pairs of r values for Na with each other variable are significantly different. The slopes of linear-regression lines for each element using weight percentages and g/100cc *appear* to be comparable, but with lines for the latter being consistently steeper than those for the former. For example, on a Harker diagram for Na vs. Si, the slope is steeper for g/100cc ($b = -0.032$, $r = -0.422$) than for weight percentages ($b = -0.015$, $r = -0.277$). Similarly, for Lakeview Mountains Pluton, Na_2O vs. SiO_2 linear-regression-line slope is -0.029 ($r = -0.408$) for weight percentage and -0.050 ($r = -0.530$) for g/100cc data. When data for several units are plotted, subtle but possibly important regression-line pattern differences occur (Fig. 1). Because Harker-diagram slope differences (e.g., in granite-suite identification, Chappell, 1984) and data fields (e.g., in trace-element discrimination diagrams for

Table 6: Correlation-Coefficient Matrices and Selected Regression-Line Slopes for wt% and g/100cc Data, San Gabriel Mountains Granitoid Rocks

tan θ*			−0.219	−0.356	−0.187	−0.312	−0.015	0.134	−0.044
					Weight percentages				
		Si	Al	Fe	Mg	Ca	Na	K	Ti
	Si		−0.904	−0.984	−0.947	−0.980	−0.277	0.810	−0.924
−0.330	Al	−0.883		0.843	0.759	0.866	0.355	−0.772	0.803
−0.457	Fe	−0.971	0.874		0.959	0.971	0.201	−0.826	0.925
−0.241	Mg	−0.944	0.794	0.961		0.951	0.129	−0.809	0.869
−0.400	Ca	−0.964	0.893	0.973	0.952		0.209	−0.871	0.928
−0.032	Na	−0.422	0.529	0.390	0.309	0.396		−0.242	0.252
0.150	K	0.764	−0.773	−0.811	−0.798	−0.857	−0.383		−0.816
−0.057	Ti	−0.908	0.834	0.929	0.876	0.932	0.427	−0.804	
					Grams/100 cc				

*tan θ = Tangent of angle of slope of regression line of each oxide with Si (*i.e.*, b of equations such as Na = $a + b \cdot$ Si)

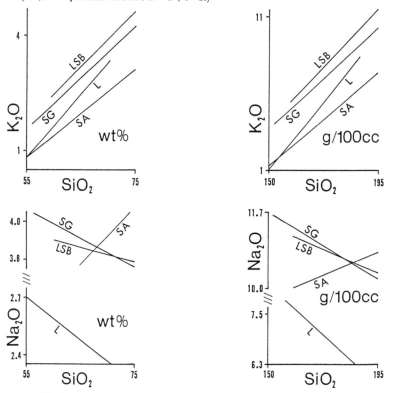

Figure 1: Harker-type diagrams for Lakeview Mountains Pluton (L), San Gabriel Mountains (SG), Little San Bernardino Mountains (LSB), and Santa Ana Mountains (SA) granitoid wt% and g/100cc data (represented by linear regression lines).

tectonic interpretation, Pearce *et al.*, 1984) are claimed to have great petrogenetical significance, the dissimilar patterns for wt% and g/100cc data could be important. It is impossible to anticipate the Harker-diagram slope differences between wt% and g/100cc data without accurate whole-rock specific gravity for each sample.

To plot open g/100cc data on a ternary diagram involves converting some variables to proportions of a constant sum (percentages), thereby introducing the standard severe problems associated with attempts to identify on ternary diagrams trends and patterns in closed data.

Because of closure constraints inherent in wt% data, and to avoid the need for data transformations (such as log-ratios or the addition of white noise), weight-per-unit volume should be explored in most petrological applications. Oxide weight percentages and g/100cc data for a rock assemblage comprise sets of different variables, so they will yield differences when used to discern patterns and interrelationships between sampled populations.

Acknowledgments

Prof. William B. Wadsworth gave valuable help in retrieving the original Lakeview Mountains Pluton data and Dr. Maurice Stone made useful comments on an early draft of this paper, which is based on an oral presentation at the joint meeting of the Royal Statistical Society/Geosciences Information Group (Geological Society of London) held November 7, 1991, in Edinburgh.

References

Aitchison, J., 1982, The statistical analysis of compositional data: *Jour. Royal Statistical Soc.,* v. 44B, p. 139–177.

Aitchison, J., 1986, *The Statistical Analysis of Compositional Data:* Chapman & Hall, London, 416 pp.

Aitchison, J., 1990, Relative variation diagrams for describing patterns of compositional variability: *Math. Geol.,* v. 22, p. 487–511.

Baird, A.K., Baird, K.W. and Welday, E.E., 1979, Batholithic rocks of the northern Peninsula and Transverse Ranges, southern California: Chemical composition and variation, *in* Abbott, P.L. and Todd, V.R., (eds.), *Mesozoic Crystalline Rocks: Peninsula Ranges Batholith and Pegmatites Point Sal Ophiolite:* San Diego State Univ., Dept. Geol. Sci., San Diego, California, p. 111–132.

Baird, A.K., McIntyre, D.B. and McColl, R., 1962, A test of the precision and sources of error in quantitative analysis in light, major elements in granitic rocks by X-ray spectrography: *Advances in X-ray Analysis,* v. 5, p. 412–422.

Baird, A.K., McIntyre, D.B. and Welday, E.E., 1967*a*, Geochemical and structural studies in batholithic rocks of southern California, Part II: Sampling of the Rattlesnake Mountain Pluton for chemical composition, variability, and trend analysis: *Geol. Soc. Amer. Bull.,* v. 78, p. 191–222.

Baird, A.K., McIntyre, D.B., Welday, E.E. and Morton, D.M., 1967*b*, *A Test of Chemical Variability and Field Sampling Methods, Lakeview Mountains, Southern California Batholith:* California Div. Mines and Geology, Special Report 92, p. 11–19.

Boyle, R.W., 1961, *The Geology, Geochemistry, and Origin of the Gold Deposits of the Yellowknife District:* Geol. Survey of Canada, Memoir 310, p. 1–193.

Chappell, B.W., 1984, Source rocks of I- and S-type granites in the Lachlan Fold Belt, southeastern Australia: *Philosophical Trans. Royal Soc.*, ser. A, v. 310, p. 693–707.

Chayes, F., 1948, A petrographic criterion for the possible replacement of rocks: *Am. Jour. of Science*, v. 246, p. 413–425.

Chayes, F., 1960, On correlation between variables of constant sum: *Jour. Geophys. Research*, v. 65, p. 4185–4193.

Chayes, F., 1971, *Ratio Correlation: A Manual for Students of Petrology and Geochemistry:* Univ. Chicago Press, Chicago, 99 pp.

Chayes, F., 1983, On the possible significance of strong positive correlations between trace elements: Unpubl. typescript, Geophysical Laboratory, Washington, D.C., 6 pp.

Exley, C.S., 1959, Magmatic differentiation and alteration in the St. Austell Granite: *Geol. Soc. London Quart. Jour.*, v. 114 [for 1958], p. 197–230.

McIntyre, D.B., Welday, E.E. and Baird, A.K., 1965, Geologic application of air-comparison pycnometer: A study of the precision of measurement: *Geol. Soc. Amer. Bull.*, v. 76, p. 1055-1060.

Morton, D.M., Baird, A.K. and Baird, K.W., 1969, The Lakeview Mountains Pluton, southern California batholith, Part II: Chemical composition and variation: *Geol. Soc. Amer. Bull.*, v. 80, p. 1553–1564.

Pearce, J.A., Harris, N.B.W. and Trindle, A.G., 1984, Trace element discrimination diagrams for the tectonic interpretation of granitic rocks: *Jour. Petrology*, v. 25, p. 956–983.

Rock, N.M.S., 1988, *Numerical Geology: A Source Guide, Glossary and Selective Bibliography to Geological Uses of Computers and Statistics:* Springer–Verlag, Berlin, 427 pp.

Sarmanov, O.V. and Vistelius, A.B., 1959, On the correlation of percentage values: *Doklady Akademii Nauk SSSR*, v. 126, p. 22–25.

Sharp, W.E., 1989, Opening the closure problem: *Geol. Soc. Amer. Abstracts with Programs*, v. 21, p. A58.

Stanley, C.R., 1990, Descriptive statistics for N-dimensional closed arrays: A spherical coordinate approach: *Math. Geol.*, v. 22, p. 933–956.

Tyrrell, G.W., 1948, A boring through the Lugar Sill: *Geol. Soc. Glasgow Trans.*, v. 21, p. 157–202.

Tyrrell, G.W., 1952, A second boring through the Lugar Sill: *Geol. Soc. Edinburgh Trans.*, v. 15, p. 374–392.

Wager, L.R., 1929, Metasomatism in the Whin Sill of the north of England, Part I: Metasomatism by lead vein solutions: *Geological Magazine*, v. 64, p. 97–110.

Whitten, E.H.T., 1972, Enigmas in assessing the composition of a rock unit: A case history based on the Malsburg Granite, SW. Germany: *Geol. Soc. Finland Bull.*, v. 44, p. 47–82.

Whitten, E.H.T., 1975, *Appropriate units for expressing chemical composition of igneous rocks:* Geol. Soc. Amer., Memoir 142, p. 283–308.

Whitten, E.H.T., 1987, Classification and partitioning of igneous rocks, *in* Prohorov, Yu. V. and Sazonov, V.V., (eds.), *Proceedings 1st World Congress Bernoulli Society (International Statistical Institute), [Tashkent, USSR, 1986]:* International Science Publ., Utrecht, Netherlands, v. 2, p. 573–577.

Whitten, E.H.T., Bornhorst, T.J., Li, G., Hicks, D.L. and Beckwith, J.P., 1987, Suites, subdivision of batholiths, and igneous-rock classification: Geological and mathematical conceptualization: *Am. Jour. of Science*, v. 287, p. 332–352.

Woronow, A. and Love, K.M., 1990, Quantifying and testing differences among means of compositional data suites: *Math. Geol.*, v. 22, p. 837–852.

Zhou, D., Chen, H. and Lou, Y., 1991, The logratio approach to the classification of modern sediments and sedimentary environments in northern South China Sea: *Math. Geol.*, v. 23, p. 157–165.

17

AMPLITUDE AND PHASE IN MAP AND IMAGE ENHANCEMENT

Joseph E. Robinson

Many geographic information systems (GIS) and computer mapping packages for graphic display provide data enhancement routines that utilize spatial filtering and may even include sample filter operators. With care, spatial filtering or convolving a filter with string or gridded data can be very effective in enhancing digital information. Filtering is intended to highlight specific features, remove unwanted noise or restore an image. The filtered output may be smoothed, it may portray features of a specific size, or it may subject data to mathematical operations. Results depend on the form of the filter operator. Fourier analysis of the filter operator produces frequency and phase diagrams that describe the effect the filter will have on the input data. These diagrams indicate how the component frequencies in the input data will be altered by the filter. Phase relates to the position of contained features, and amplitudes are responsible for feature definition. Derivative filtering may alter both phase and amplitude and completely change the appearance of the output. Spatial filtering can be very effective, however the interpreter must exercise great care selecting suitable filters. Fourier analysis can aid in design of a suitable filter and in the resultant evaluation. Examples herein describe some common three-value filters for string data and three-by-three matrix filters for grid and raster data.

Spatial Filters

Digital spatial filters are arrays of values with the same form and interval as the original numeric data set (*e.g.*, Robinson and Treitel, 1969). Filters may be strings of one-dimensional values, two-dimensional gridded data sets, or even multi-dimensional data. Spatial filters to be used with digital data sets can be designed and evaluated in either the spatial or the frequency domain. Similarly, line, map, and image data can be transformed and viewed in both domains. Fourier analysis is the bridge between the two domains (Cooley and Tukey, 1965; Gentleman and Sande, 1966; Robinson and Cohn, 1979; Sondegard, Robinson and Merriam, 1989). The filtering operation can be carried out in either domain by convolution in the spatial domain or by multiplication in the frequency domain. Convolution is the Fourier transform of multiplication. Although spatial filters are usually designed in the spatial domain, their effect on the input information is most easily evaluated in terms of frequency (*e.g.*, Blackman and Tukey, 1958; Otnes and Enockhson, 1972). In this domain the amplitude spectrum, which relates to the amplitudes and definition of the component frequencies in the data, can be visualized by multiplying the amplitude of the input components by the amplitude of the corresponding filter components while the phase of the filter components is added to that of the input components. Phases with values other than zero alter the form and position of features in the output. All spatial data can be described in terms of their component frequencies. In general, the long-wavelength, low spatial frequencies make up the large regional features. The middle frequencies are represented by the intermediate features and the short-wavelength high frequencies by the small features and random noise. The high frequencies are also important for giving form to the shapes and boundaries of all features.

The frequency content of the input data can be determined from its Fourier transform. Frequencies can range from zero to a maximum which has a wavelength not less than twice the distance between grid or pixel intervals (Shannon, 1948). Unwanted information or noise forms an added component to desired frequencies. The effect of noise can be reduced by using filters to alter the signal-to-noise ratio in the affected frequencies. Random noise and atmospheric or path noise in images normally affect the highest frequencies and may be accentuated by image enhancement methods that amplify these high ranges. It may be necessary to cascade different enhancement filters to obtain the desired results.

Spatial filters usually fall into three general categories: low-pass for overall smoothing of regional features, band-pass for isolating intermediate features, and high-pass for accentuating small features, as well as for increasing definition of the shape and form of all classes of features. In addition, there are deconvolution filters which attempt to shape the entire spectrum to an ideal form and edge or boundary detectors that are useful in pattern recognition. There is also a special filter, the all-pass filter, which can be useful for constructing other filters and for adjusting amplitude of the data set. Filters can be added and subtracted in the same way as other mathematical functions. They can be convolved with each other in the spatial domain and multiplied in the frequency domain. The average value in any

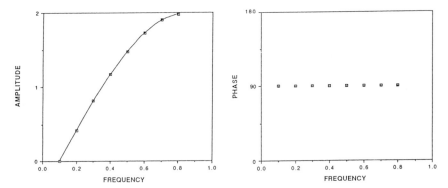

Figure 1: Amplitude and phase spectra of first-derivative operator $(1,-1)$. The phase spectrum indicates a constant 90° phase shift, so slopes in the output will be positive or negative features according to their direction.

data set is the zero frequency and a spatial filter whose weights sum to unity will not change this average value. Similarly, if the sum of the filter weights is equal to zero, the average value of the input data is removed and the amplitude of the zero frequency is reduced to zero. The average value and the range of values in the filtered output is controlled by adjusting the sum of the filter values. However, band-pass and high-pass filters do not pass the zero frequency and the sum of the filter weights must equal zero. In the latter cases, amplification is controlled by altering the values of the filter weights without changing their sum. The effect on individual frequencies can be seen in the amplitude spectrum of the Fourier transform. As long as the filters have radial symmetry and a zero-phase spectrum, filtering does not alter the position of any feature.

Spatial Filter Construction

A practical method for the design, construction, and evaluation of a spatial filter can be demonstrated through the development of a simple derivative filter. The first derivative of any function is its slope. The output is zero for any set of constant values, positive for an increasing set, and negative for a decreasing set. The average value is deleted and the sum of the filter weights must equal zero. For string data the digital filter operator is $(1,-1)$. Convolution entails the reversing of the filter operator, then multiplying and summing filter and input data values as the filter is moved along the data set. The Fourier transform spectra of this filter (Fig. 1) indicate the amplitudes of the frequency components range from zero at the low frequencies and smoothly increase to a maximum at the highest frequency.

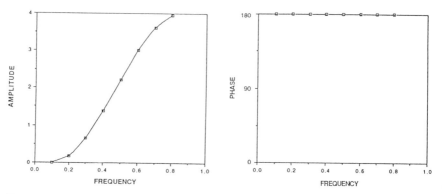

Figure 2:   Amplitude and phase spectra of second-derivative operator (1,−2,1) obtained by convolving first-derivative operator with itself. The output describes rate of change of slope, so filtered features are sharper but inverted.

The phase diagram displays a constant 90° phase shift causing the filtered output to appear as positive and negative slopes. In this case, the amplitude diagram has the form of the first quadrant of a sine wave and the filtered output will be the best digital approximation of a first-derivative function.

   The second derivative displays the rate of change of the slope of the data and can be obtained by convolving a second time with the first-derivative operator or by a single convolution with a second-derivative operator. The one-dimensional second derivative is obtained by convolving the first-derivative operator with itself. This second-derivative operator is (1,−2,1) and its frequency transform is displayed in Figure 2. This filter narrows and sharpens features displayed in the data but inverts all undulations, as is indicated by the 180° phase shift. The inversion problem can be circumvented by reversing one of the first-derivative operators. In this latter case the two 90° phase shifts cancel and the new operator has zero phase. The second-derivative operator now is (−1,2,−1) and its frequency diagrams are displayed in Figure 3. The amplitude characteristics are the same as in the previous filter, however, it now has zero-phase characteristics which will ensure the positional validity of displayed features. This operator often is applied to enhance the definition of features in magnetic and gravity maps and to improve remotely sensed images. As long as the filter values are changed in the same proportion, the filter will have the same characteristics, however, the amplitude of the filtered features will be altered. A commonly used version of this operator is (−0.5,1,−0.5), although for abnormally low or high amplitude data the filter values can be changed to improve the amplitudes of the filtered features. The symmetry of the filter values reflects the zero-phase properties of the phase spectrum.

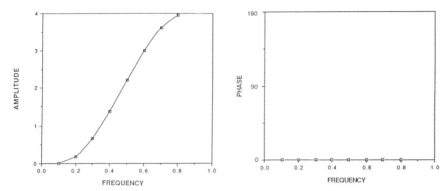

Figure 3: Amplitude and phase spectra of zero-phase second-derivative operator $(-1,2,-1)$ obtained by convolving a reversed first derivative with a normal one. Filtered features are displayed in their correct attitudes.

With gridded or raster image data the one-dimensional derivative filter can first be applied row by row then column by column. The application of a simple row-by-row convolution only filters in one of the two dimensions and does not completely filter the data set. However, it is possible to filter the full data set in one pass by generating a two-dimensional derivative filter operator. This two-dimensional filter can be constructed by convolving the one-dimensional filter with itself, row by column, to produce a three-by-three matrix; for example $(0.25,-0.5,0.25; -0.5,1,-0.5; 0.25,-0.5,0.25)$. If the zero-phase one-dimensional filter is used for the construction, the two-dimensional form will approximate radial symmetry and have zero-phase characteristics. Amplitude can be adjusted by multiplying all the filter weights by a suitable amount. Three- or more dimensional filters can also be produced by continuing the convolution of the original second-derivative filter in each of the new dimensions.

Second-derivative filters are high-pass operators and are employed to increase resolution of features portrayed in the input data. However, because they pass any high frequencies that are contained in the input data, random noise components are also increased and there may be a problem with noisy image data. Where desirable, the high-frequency components can be amplified or decreased by multiplying all filter values by an appropriate factor less than or greater than one. Also, because these filters produce output on a break in slope, they are a simple form of edge detector.

Special-Purpose Filters

Smoothing filters are low-pass filters. Depending on the filter weights, they eliminate the highest frequencies and may alter the amplitude of some of the lower frequency components. The simplest form is the digital filter version of a running average. The one-dimensional, three-value, averaging-operator values are a string of ones divided by three. Similarly, the two-dimensional filter values are all equal to one-ninth. The Fourier transform of a running average has the mathematical form of a sine X over X figure as is illustrated by the amplitude spectrum in Figure 4. The phase spectrum indicates 180°

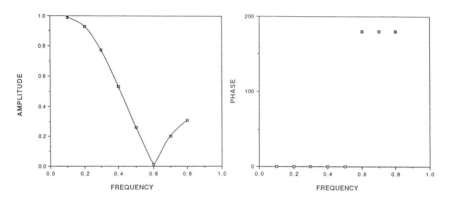

Figure 4:   Amplitude and phase spectra of running average operator (0.33,0.33,0.33). Frequency domain transform is a $\sin X/X$ function; there is a 180° phase inversion in some high frequencies.

changes in higher frequencies, signaling an inversion in these components and in any filtered features they represent. This phase inversion could affect the display of important small, but subtle features and where this is important, a useful smoothing filter can be obtained by applying a triangular weighting to the filter values. The closest approximation of a triangular weighting in a three-point string filter is to have both side values equal to one-half the center value. If the sum of the filter values is equal to one, the average value of the input data is unchanged; if more than one, it is increased and if less than one, it is decreased. The Fourier transform of the triangular filter is displayed in Figure 5. It should be noted that the rate of attenuation from low to high frequencies is considerably less than for the previous filter. Filter functions with the shape of a Gaussian curve are even better smoothing filters but require more than three values for effective definition.

High-pass filters retain only the high-frequency component. They include derivative filters and edge detectors. The sum of the filter weights must equal zero to completely eliminate the zero frequency. Possibly the easiest way to manufacture a high-pass filter is to subtract a low-pass filter from an all-pass operator. Figure 6 illustrates the spectra obtained by subtracting the low-pass filter of Figure 5 from an all-pass filter. The amplitude of the retained frequencies can be adjusted by multiplying all filter weights by a suitable

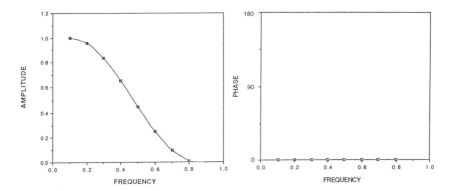

Figure 5: Amplitude and phase spectra of weighted running-average operator (0.25,0.5,0.25). Symmetrical weighting approximates a triangle; there are zero-phase characteristics over the full spectrum.

constant which must be the same for all weights in order to maintain the zero sum.

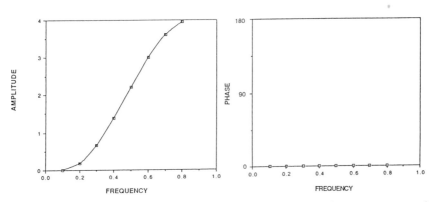

Figure 6: Amplitude and phase spectra of high-pass filter (−0.25,0.5, −0.25) obtained by subtracting operator in Figure 5 from all-pass operator (0,1,0).

Band-pass filters are intended to eliminate both low- and high-frequency components in the input data in order to display features of specific size. When the filter is limited to a three-value set it is not possible to construct an ideal filter of this type because additional values are required to provide the necessary definition. One possible filter that passes a central range of frequencies is (1,0,−1). Unfortunately, as illustrated in Figure 7, this filter has a 90° phase shift and would require a second pass with the filter reversed to correct the phase.

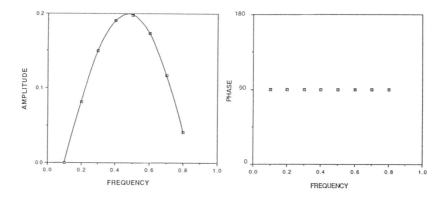

Figure 7: Amplitude and phase spectra of simple band-pass operator (1,0,−1) constructed with a three-value array. The 90° phase shift can be converted to zero phase by a second pass with filter reversed.

Deconvolution filters are designed to convert the input data to an optimum frequency spectrum. With image data, deconvolution filters are used to correct for atmospheric and path distortion of the recorded data. In this case a great deal can be accomplished with a three-value string or a three-by-three matrix. Most degradation of image data results from attenuation of the higher frequency components and can be corrected by filters that pass all frequencies but increase amplification of the high-frequency components. The amplitude spectra of some filters that pass all frequencies with different rates of gain from the low to the high frequencies are illustrated in Figure 8. These filters have the same basic form and their weights can be adjusted to achieve a suitable rate of gain in the high frequencies while retaining a range of amplitudes for the lower frequencies. All are zero-phase filters.

Each map or image will have a theoretically ideal, possibly unique frequency spectra which could only be realized in a perfect, noise-free system. However, a reasonable estimate of the ideal spectrum can be determined by considering spectra from similar terrain that have been sampled under very good conditions. The Fourier transform of even a single raster across an image taken from a low-flying aircraft or a ground receiver would be ideal. A best estimate based on experience may be suitable for choosing a filter. Effective deconvolution filters can be constructed by picking the filter spectra that best converts the spectra of the degraded image to the ideal. The best choice is the amplitude spectra of a filter that when multiplied by the map or image spectra best approximates the ideal. However, three-value filters are very limited and it is usually only effective to make modest adjustments in gain rates. Again, zero-phase filters are important for feature stability.

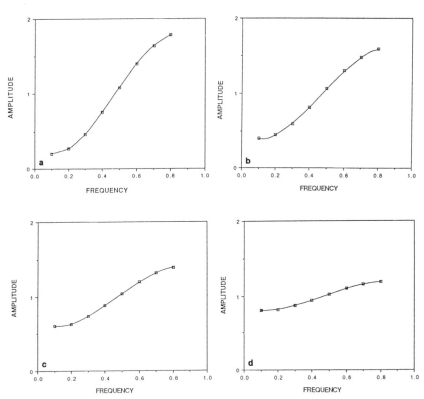

Figure 8: Amplitude spectra of four deconvolution operators with different rates of gain. All have zero-phase characteristics. [a] $(-0.4,1,-0.4)$ has highest gain, [b] $(-0.3,1,-0.3)$ and [c] $(-0.2,1,-0.2)$ moderate gain, and [d] $(-0.1,1,-0.1)$ low gain. Deconvolution filters can be evaluated by multiplying amplitude spectra by object function.

Conclusions

Fourier transforms provide an approach to evaluating the properties of spatial data and spatial filters. The amplitudes of the frequency components in the filter when multiplied by the corresponding amplitudes in the input function will indicate the results in the filtered output. The phase values of the filter are added to those of the input to give the output phase. With both filter and input function in the frequency domain the best filter can be determined. Filter properties can be used to shape the spectra and thus enhance the presentation of features present in the input information. When judiciously designed and correctly applied, spatial filters are a very effective tool in data enhancement and analysis. Although most spatial data are in two dimensions, spatial filter design can be visualized and carried out initially in one dimension for convenience then expanded to a two-dimensional form. Two- or more dimensional filters can all be constructed by convolving a one-dimensional row filter with each column of each of the new dimensions.

References

Blackman, R.B. and Tukey, J.W., 1958, *The Measurement of Power Spectra:* Dover Publications, New York, 190 pp.

Cooley, J.W. and Tukey, J.W., 1965, An algorithm for the machine calculation of complex Fourier series: *Math. Computing,* v. 19, p. 297–301.

Gentleman, W.M. and Sande, G., 1966, Fast Fourier transforms—For fun and profit: *Proc. AFIPS Conference,* v. 29, p. 563–578.

Otnes, R.K. and Enockhson, L., 1972, *Digital Time Series Analysis:* John Wiley & Sons, New York, 467 pp.

Robinson, E.A. and Treitel, S., 1969, *The Robinson-Treitel Reader:* Seismograph Service Corp., Tulsa, Okla., 225 pp.

Robinson, J.E. and Cohn, B.P., 1979, FILTRAN; A FORTRAN program for one-dimensional Fourier transforms: *Computers & Geosci.,* v. 5, no. 2, p. 231–249.

Shannon, C.E., 1948, A mathematical theory of communication: *Bell System Tech. Jour.,* v. 27, p. 379–423, 623, 656.

Sondegard, M.A., Robinson, J.E. and Merriam, D.F., 1989, *Filt-PC, A One-Dimensional Fast Fourier Program in FORTRAN for PC's:* Microcomputers in Geology 11, Pergamon Press, Oxford, p. 251–259.

18

FRACTALS IN GEOSCIENCES—CHALLENGES AND CONCERNS

Ute Christina Herzfeld

Two alternative "definitions" of self-similarity:

So Nat'ralists observe, a Flea
Hath smaller Fleas that on him prey;
And these have smaller Fleas to bite 'em;
And so proceed ad infinitum.   —Jonathan Swift

In every drop of dew is an entire
world with its own drops
of dew containing
a world.   —Leibniz

"Fractals" and "chaos" have become increasingly popular in geology; however, the use of "fractal" methods is mostly limited to simple cases of self-similarity, often taken as the prototype of a scaling property if not mistaken as equivalent to a fractal as such. Here a few principles of fractal and chaos theory are clarified, an overview of geoscience applications is given, and possible pitfalls are discussed. An example from seafloor topography relates fractal dimension, self-similarity, and multifractal cascade scaling to traditional geostatistical and statistical concepts. While the seafloor has neither self-similar nor cascade scaling behavior, methods developed in the course of "fractal analysis" provide ways to quantitatively describe variability in spatial structures across scales and yield geologically meaningful results.

The Realm of the Appleman

Upon hearing the slogan "the appleman reigns between order and chaos" in the early 1980's and seeing colorful computer-generated pictures, one was simply fascinated by the strangely beautiful figure of the "appleman" that, when viewed through a magnifying glass, has lots of parts that are smaller, and smaller, and smaller applemen. The "appleman" is the recurrent feature of the Mandelbrot set, a self-similar fractal, and in a certain sense, the universal fractal (*e.g.*, see Peitgen and Saupe, 1988, p. 195 *ff*.). Soon the realm of the appleman expanded, made possible by increasing availability of fast, cheap computer power and increasingly sophisticated computer graphics.

In its first phase of popularity, when the Bremen working group traveled with their computer graphics display seeking public recognition through exhibits in the foyers of savings banks, the fractal was generally considered to be a contribution to modern art (Peitgen and Richter, *The Beauty of Fractals*, 1986). While the very title of Mandelbrot's famous book, *The Fractal Geometry of Nature* (1983), proclaims the discovery of the *proper* geometry to describe nature, long hidden by principals of Euclidean geometry, the "fractal" did not appeal to Earth scientists for well over two decades after its rediscovery by Mandelbrot (1964, 1965, 1967, 1974, 1975).

By the mid-1980's fractal geometry had been applied to non-linear physics (Grassberger and Procaccia, 1983), meteorology (Lovejoy and Mandelbrot, 1985), and astrophysics. Still in 1986, when I proposed to use fractal processes to characterize the structure of coal mines, I must speak of "scaled variography" to be accepted by geologists. Mention of "deterministic chaos" in a geology seminar provoked bursts of laughter. Time has brought about a change, almost to the opposite extreme. It is now fashionable to talk about fractals whenever possible and search for self-similarity in all kinds of geological features. The quality of science using fractal concepts in geology varies broadly. An attempt is made here to clarify some principles and terminology. Basic questions will be: Are methods being applied correctly? What can we learn about the Earth from "fractal" analysis?

Concepts and Definitions

A *fractal* is, by definition, an object of Hausdorff–Besicovitch dimension strictly exceeding its topological dimension. This usually implies that the dimension is non-integer. Exceptions include the space-filling Peano curve with Hausdorff dimension 2 and topological dimension 1 and Cantor's devil's staircase (*cf.* Mandelbrot 1983, plate 83) that has both Hausdorff dimension 1 and topological dimension 1, but should intuitively be called a fractal. [More mathematically stated definitions can be found in Mandelbrot (1983, p. 349–390), Herzfeld *et al.* (1992), and Olea (1991).]

Despite only recent popularity, fractal sets are not a recent discovery. In the 19*th* century the idea that "the general notion of volume or magnitude is indispensable in investigations on the dimensions of continuous sets" was quoted by Cantor (1883). Peano (1890), Fricke and Klein (1897), Lebesque (1903) and Julia (1918) also described measures and sets that are now called "fractal" (*e.g.*, Cantor set, Peano curve, Julia dragon). The notion of fractal dimension goes back to Caratheodory (1914) and Hausdorff (1919). Besicovitch extended the definition to non-integer and non-standard shapes. The set named for Benoit Mandelbrot is the complex values c for which iterations of the map $x \rightarrow x^2 + c$ do not go to infinity for $x = 0$. The set assumes different shapes for different values of c. Its general shape was known long ago, but only by computer graphics is it possible to display it in detail.

The classic example of a line having fractal dimension stems from topography (Mandelbrot, 1967). If the coastline of Britain is repeatedly measured using sticks of decreasing length, the total length of the coastline increases. As this process does not converge to an integer, the coastline has fractal Hausdorff dimension. The roughness of the coastline displays variability at small scale. The geomorphologist's eye is captured by pictures of "Brownian landscapes" (Mandelbrot, 1983, pp. 264–265, C10–13), generated from realizations of Brownian processes. The simulations resemble slightly bizarre mountainous landscapes (and several years later motivated scientists to explore them to describe the Earth; see section on *Fractals Everywhere?*). However, Mandelbrot's Brownian landscapes not only have fractal dimensions, but are also *self-affine*. Self-affinity is a weaker form of self-similarity.

An (unbounded) set is *self-similar* if part of it looks like the whole set at a suitable enlargement (the *scale ratio*), up to displacement and rotation. An example of a deterministic self-similar fractal, generated in a step-by-step process, is the space-filling Peano curve. If an affine transformation can be used to map a set onto a part of itself, the set is called "self-affine," (*i.e.*, different scale ratios can be used along coordinate axes in the transformation, otherwise the definition is the same as for "self-similarity"). Geometrical step-by-step processes usually generate deterministic fractals, while in the Earth sciences, simulations of stochastic processes are more common. The Mandelbrot landscape is a realization of a stochastic fractal.

It is a widespread error to think that the terms "fractal" and "self-similar" are synonymous. Many fractal sets are neither self-similar nor self-affine; the straight line, for example, is self-similar, but no one would call it a fractal (its topological dimension and Hausdorff dimension are both 1). This error is at the basis of many "applications" of fractal analyses in geology, as discussed in the section on *Possible Pitfalls*. The *similarity dimension D* of an irregular line can be estimated. Let L_i, $i = 1, \ldots, m$, with m a natural number, be the total length measured using a ruler of length r_i in a survey using at least two rulers of different sizes. Then the slope of the regression line of $(\log L_i)$ versus $(\log r_i)$ is equal to $1 - D$. The similarity dimension of an irregular surface can be estimated in the same manner. It is often called the "fractal dimension" and used to estimate the Hausdorff dimension of a geological object—usually without questioning whether the object under study is at all self-similar!

A term often found with "fractal" is "chaos." These two concepts are rooted in different branches of science. Chaos is the state of a dynamic system in which long-term prediction is impossible. Dimension concepts are part of topology and function theory; "fractal geometry" is a geometrical concept. Introductory books on fractals and chaos are abundant and generally fall into two classes. Graphically oriented books on fractals typically feature a geometrical approach involving programming and simulation of pictures (*e.g.*, Peitgen and Richter, 1986; Peitgen and Saupe 1988). Books on introductory chaos theory focus on bifurcations and chaotic behavior in dynamical systems (*e.g.*, Holden, 1986). Mathematical topics that provide theory related to fractals and chaos are function theory, complex analysis, topology (related mostly to fractals), differential equations, dynamical systems, and bifurcation theory (related to chaos theory), and stochastic processes and spatial statistics, relating to both fractals and chaos.

Fractals Everywhere?—A Short Review

Recent years have seen an explosion of fractal- and chaos-related essays in the Earth sciences. Following is a summary of some of the work according to field of application and mathematical concept with respect to method of analysis or simulation: data analysis *vs.* simulation *vs.* prediction and deterministic *vs.* probabilistic approach.

Field of Application. Peitgen and Richter (1985) observed a Mandelbrot set studying phase-transitions of ferromagnetic systems in computer simulations, which gave rise to their computer graphics. Climate research was the first Earth science discipline to explore chaotic systems, perhaps because climate is closely related to meteorology (rain fields; *e.g.*, Lovejoy and Mandelbrot, 1985) and astrophysics (sunspot cycles; *e.g.*, Mundt *et al.*, 1990), where concepts of chaos and fractals were first employed. Studies which consider climate as driven by a chaotic dynamic system can be ordered according to time scales: (a) interannual and (b) glacial/interglacial. Studies on an interannual scale include prediction of El Ninos and Southern Oscillation (*e.g.*, Krishnamurthy, 1990). Climatic changes on a glacial/interglacial scale have been investigated by Nicolis and Nicolis (1985), Ghil (1990), and Matteucci (1990). Only limited time ranges can be studied because of limitation of available data, which are either weather recordings over past years, sediment cores, or geological records from ice cores; in each case, data have a specific range of coverage and a specific resolution.

Seismologists have attempted to describe crust formation by fractal models and use these to predict earthquakes. Since complexity of fault patterns or statistical occurrence of earthquakes cannot be explained by simple models, descriptions based on chaotic behavior or fractal dimension have been sought (Turcotte, 1986; Bak, 1990; Brown *et al.*, 1990; Huang, 1990; Steacy and Sammis, 1990; Kagan, 1991; Varnes, 1991). In most instances, calculations are based on a spring block model. Often simulations rather than data analyses support the rationale for the fractal model.

Another large group favoring fractal concepts are physical geographers who quantitatively analyze topography at various scales (see Sayles and Thomas, 1978; Burrough, 1981; Goodchild, 1988; Culling, 1989; Elliot, 1989; Jones *et al.*, 1989; Unwin, 1989). The usefulness of fractal concepts for describing seafloor topography is discussed by Bell (1975), Malinverno (1989),

Goff *et al.* (1991), and Herzfeld *et al.* (1992). Smith and Shaw (1990) explore the idea that seafloor topography is a product of a large system governed by a chaotic dynamic system, but present no evidence for the existence of a strange attractor in their analysis. There are also individual studies of other geological features. Goodchild (1988) studies lakes, Nicholl (1990) geysers, and Stewart (1990) mantle convections.

Mathematical Method Applied. Nicolis and Nicolis (1985), based on an analysis of core data, questioned whether there is a climate attractor. Since then, studies in climate dynamics use the dynamical system approach, searching for descriptions that use free or forced oscillators (Ghil, 1990), basins of attractors (Tsonis and Elsner, 1990), and Lyapunov exponents. These works are based mostly on deterministic chaos. Inspired by pictures of "fractal landscapes" in Mandelbrot's book, generalized Brownian noise has been favored as an approximation of terrestrial or marine topography (*e.g.*, Bell, 1975; Sayles and Thomas, 1978, Jones *et al.*, 1989, Malinverno, 1989). As noted previously, Brownian landscapes are self-affine. On the other hand, spectral analyses of topography that contradict the scaling property of Brownian noise have been carried out by Berkson and Matthews (1983), Fox and Hayes (1985), and Brown and Scholz (1985). Power law models of seafloor topography are not consistent with Mandelbrot's Brownian landscapes. Data analysis methods used to study seafloor topography are mostly spectral, and stochastic for simulations (*e.g.*, Goff *et al.*, 1991). Geostatistical methods are used by Chilès (1988), Elliot (1989), and Herzfeld *et al.* (1992).

Possible Pitfalls: Data Ranges and Assumptions

There are several sources of problems with fractal analyses. First, there is no unique definition of the fractal throughout the literature, as noted earlier. Geological and geographical studies often consist only of an estimation of the similarity dimension from a log–log plot to which a line is fitted (see section on *Concepts and Definitions*). This derivation of the dimension is usually mistaken as a proof that the object under study is fractal or self-similar (often the terms are not distinguished).

A primary problem with observing scaling properties from data sets is that the range of scales covered by the data is often very limited. In geophysical surveys, the maximal resolution is determined by the survey system, while the time necessary to sample larger areas limits the upper end of the spatial scale. As a result, a ratio of 10^2 of the largest to the smallest scale studied is seldom exceeded (*e.g.*, Chilès 1988; Mutter *et al.*, 1990). Meteorological studies of rainfields and clouds combine satellite images and millimeter-scale resolution experiments to achieve a scale ratio of 10^9 (Lovejoy and Mandelbrot, 1985; Tessier, Lovejoy and Schertzer, 1991).

Few workers have built instruments for high-resolution sampling, but Elliot (1989) describes a sampling device designed to study roughness in a glacial foreland down to very small lags (0.05 m) and Brown and Scholz (1985) designed an instrument to measure rock surfaces from millimeter to meter resolution. They found that the dimension of fractures depends on scale of resolution; it decreases with resolution from 2.5 to 2. Wong and Lin (1988) studied small-angle scattering from rock surfaces at submicron scale using X-ray techniques. The lack of adequate coverage across scales introduces "edge effects" in the estimation of parameters and hides spatial structures. Error bounds for the high- and low-resolution ends of the scale need to be calculated. Artificial scale limits induced by observation should not be confused with limits of "natural fractals" (such as a broccoli plant) that have a maximal and a minimal size between which the scaling behavior is observed.

Much present work on fractals in the geosciences involves only simulations (Turcotte, 1986; Yfantis *et al.*, 1988; Ghil, 1990; Huang, 1990; Matteucci, 1990; Goff, 1991) or fails to prove that a suggested model fits observations (Smith and Shaw, 1990). Unfortunately, some essays mention fractals in the theoretical part, but do not use them in the data analysis (*e.g.*, Malinverno and Gilbert, 1989)—an unfortunate consequence of bad style and the current fashionability of the fractal? Fox (1989) finds discrepancies between theoretically expected and empirically derived results when measuring the fractal dimension of time series with the ruler method. Contrary to predictions of Berry and Lewis (1980), fractal dimension is not linearly related to the spectral exponent. Instead, the functional form depends on resolution in relation to total length of the sequence and minimum distance between observations. In an analysis of topographic profiles from the Adirondack mountains, Hough (1990) points out that logarithmic compression of high frequencies may hide important details in the spectrum that indicate deviations from a strict fractal character.

Geostatistics and Fractals

Many problems now addressed as fractal have traditionally been approached with spectral or geostatistical methods. Spatial behavior of a variable can be characterized by the variogram or the covariance function, or by the power spectral density (for definitions, see Matheron, 1963; Fox, 1989; Journel and Huijbregts, 1989; Herzfeld, 1992). Scaling problems are typical for variography in exploration geology. The ambiguity of the random or deterministic problem, reflected in the transitional status of the regionalized variable in geostatistics, appears in fractal analyses and simulations. There are two major differences:

(a) *Estimation/Simulation*: The kriging estimator is a linear estimator (or interpolator) that simplifies structures and smooths the variability of a rough surface. Conditional simulation retains the roughness at a given scale. Fractal (sets and) processes, on the other hand, comprise many more complex structures, of which objects of integer dimension are only a special case. But most animals in the fractal zoo are not understood even in mathematical theory. At present, only the special case of the self-similar (or self-affine) process is being explored in the geosciences.

(b) *Prediction problems*: The classical question in geostatistics is: What can we learn from observations here about conditions there? Predictions are made on the same scale, but at unexplored locations. The quality of the prediction decreases with increasing distance from explored sites. The corresponding problem in "fractal" prediction is: What can we learn from observations on a small scale about conditions on a large scale, and *vice versa*? The prediction is from one scale to another, but in the same location.

These differences concern the estimation step, but not the structure analysis. Simple examples of scaling problems are those encountered by geostatisticians describing an ore deposit, say by two nested variograms with different ranges (*e.g.*, Journel and Huijbregts, 1989). Geostatistical methods have been used previously in a "fractal" context by Chilès (1988) and Elliot (1989). We propose to use geostatistical methods for "fractal" analysis (Herzfeld *et al.*, 1992). The variogram is suitable for objects that are, or might be, fractal, because it is designed to describe the spatial behavior of a regionalized variable dependent on distance and direction. The following criterion links variogram functions and self-similar/self-affine processes. It provides an easy-to-use tool in fractal analysis.

Variogram Criterion for Scaling Properties

Variogram criterion: *If a spatial (Gaussian and intrinsic) random process is self-similar, then its variogram is also self-similar.* For definitions of the variogram and intrinsic process, the reader is referred to Journel and Huijbregts (1989) and to Olea (1991). The variogram criterion is proven in Herzfeld *et al.* (1992), based on the definition of self-similarity, and after correct formulations of scaling properties for stochastic processes of the kind used in topography description (more generally, for regionalized variables) have been established. It also has been shown that if the surface is self-similar, then so is a profile (in the geologic sense, *i.e.*, a profile running across the ground or across a map). The criterion holds also for (geologically meaningful) cases of self-affinity. Turning the criterion around provides a convenient tool for determining whether a geologic object is self-similar: *Calculate the variogram at different scales. If the variogram is not self-similar (self-affine), the object is not self-similar (self-affine).*

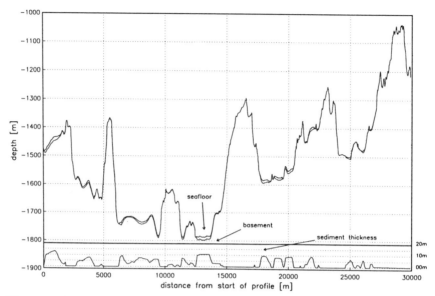

Figure 1: Deep Tow bathymetry, depth of basement and sediment thickness along a profile across the western flank of the Mid-Atlantic Ridge near 37° N.

Seafloor Topography: Scale-Dependent Structures

An example from bathymetry shows that spatial structure of the seafloor depends on scale. Data were collected with the Deep Tow system on the western flank of the Mid-Atlantic Ridge near 37° N. Deep Tow is a sonar system developed at the Marine Physical Laboratory of Scripps Institution of Oceanography. Towed by ship 50 m to 150 m above the ocean bottom, it measures depth and sediment thickness (Spiess and Lonsdale, 1982). The 30-km profile chosen for analysis (Fig. 1) runs orthogonal to the axis of the Mid-Atlantic ridge on its western flank. It was surveyed during the FAMOUS study (Macdonald and Luyendyk, 1977; Luyendyk and Macdonald, 1977) and redigitized at 5-m spacing (4097 points) by S. Miller and M. Kleinrock (pers. comm.). At large scale the profile shows a narrow abyssal hill-and-valley morphology. On various scales it seems to the eye to display the "randomness" typical of Brownian landscapes, with "a hill, on a hill, on a hill." To analyze scaling properties quantitatively, variograms were calculated at different scales of resolution (unit lag from 500 m–5 m; Fig. 2).

At a unit lag of 500 m, the variogram clearly represents the abyssal hill terrain with a spacing of ~ 6.5 km (*cf.* Macdonald and Luyendyk, 1977). At a unit lag of 125 m we see a single wavelength that is approximately linear at the origin and can be described by a spherical model up to its maximum. If the resolution is progressively increased (Fig. 2e to f), it becomes obvious that the variogram is not linear, but flattens towards the origin, indicating a

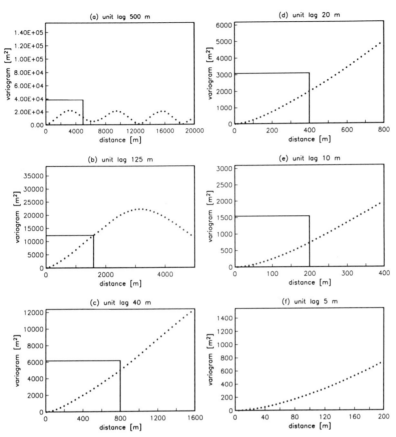

Figure 2: Variograms at different scales for Deep Tow bathymetric profile in Figure 1. Panels (b)–(f) are enlargements of near-origin area of previous panel; boxes delimit enlarged parts and indicate enlargement factor.

mean-square differentiable process. In particular, the variogram is not self-similar nor self-affine, and applying our criterion, it clearly follows that the seafloor is not self-similar/affine. Instead, it is smoother at small scale (high resolution) than at large scale. Our results are consistent with those of Fox and Hayes (1985), Brown and Scholz (1985), and Berkson and Matthews (1983) based on spectral analysis, and contradict analyses by Malinverno and Gilbert (1989) and Goff *et al.* (1991). Although the seafloor is not adequately described by a self-similar/affine process, it may still be a fractal, but formed by a more complex process with a more complex scaling behavior.

Multifractals and Cascade Processes

Rather than postulate scale invariance (self-similarity or self-affinity), Chilès (1988) described a scale-dependent similarity dimension function as a result of his study of fracture networks over a scale of 0.20 m to 20 m traced on drifts in an uranium mine located in a granite massif.

A concept of generalized scale invariance has been introduced for meteorology (Tessier, Lovejoy and Schertzer, 1991; Lovejoy and Schertzer, 1991). After finding self-similarity an unsatisfactory and too-restrictive scaling property to describe cloud fields, they defined *cascade processes*, which are multiplicative processes obeying a generalized form of scale invariance. The idea of a cascade process is that a large structure of characteristic length l_0 and density Φ_0 is broken up into substructures of characteristic length $l_1 = \frac{l_0}{\lambda}$ and density $\mu\Phi_0$, where λ and μ are scalar parameters. The terminology is inherited from turbulence theory. The scaling parameter that is a constant related to the dimension in the case of self-similar fractals (now called "monofractals") becomes a function for cascade processes. Objects considered to be generated by a cascade process are called *multifractals*.

Rather than dealing with the realization of the process (*e.g.*, the topography) itself as in self-similarity, the qth-order statistical moment of the underlying "flux" (the generating process—regardless of the question of what the "flux" generating topography might be, *e.g.*, crustal accretion or orogenesis) is described by $\lambda^{K(q)}$, where $K(q)$ is the scaling exponent. The concept is more complicated than described here, and comprises more strange animals from the fractal zoo (*cf.* Lavallée *et al.*, 1991). Matters become simpler with the observation that "most" such processes fall into a universality class of "canonical cascade processes" (Lovejoy and Schertzer, 1991). A theorem states that any such canonical process has the nice property to be characterized by only two parameters. If only canonical cascade processes are considered, a log–log plotting technique can be employed, as in self-similarity questions. The log of the scaling exponent is plotted versus the log of an exponent, and the log of the trace moment is plotted versus the log of λ. If a straight line can be fitted to a scatter plot it indicates that the process is a cascade process, and the object studied is a multifractal. Indeed, the authors claim that the scaling behavior usually found in log–log plots in geography and geology really indicates that the process is multifractal, rather than fractal and self-similar (Lavallée *et al.*, 1991).

Application of the "double-trace moment technique" of Lavallée and coworkers to Deep Tow data produced the graphs in Figure 3. Similar results were obtained for a range of parameters. Obviously, scaling behavior is not observed. Instead, independent of the parameters, there is a scale break in the plot of the trace moments at about 40 m resolution, and possibly a second one at about 640 m to 1280 m resolution (Fig. 3b and 3c). In case the log–log plot of trace moments is not linear, it is impossible to derive the scaling exponent from the graph in Figure 3(a). For a canonical multifractal, the latter should also be a straight line for a range of η-values, with the slope of that line giving the scaling exponents. To the marine geologist, the existence of scale breaks is not surprising, as bathymetric charts and bottom photographs reveal different structures at different scales. The scale breaks evident in the double-trace moment analysis match and complement the results of the scaled variography.

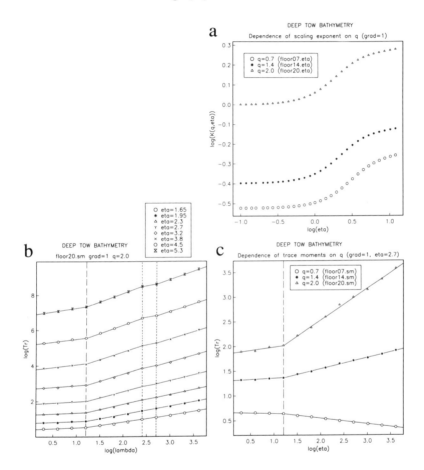

Figure 3: Multifractal analysis of Deep Tow bathymetric profile in Figure 1. (a) Dependence of scaling exponent on q. Graphs should be linear for intermediate range. (b) Log–log plot of trace moments should be linear, but scale-breaks are indicated. Dashed line corresponds to 40 m resolution, dotted lines to 640 m and 1280 m. (c) Dependence of trace moments on q.

Conclusions and Outlook

Although proven geomathematical methods generally are more widely accepted than new methods, practitioners should not hesitate to venture from the solid ground of tradition towards more adventurous concepts. A potential for analytical modeling lies in fractal geometry and fractal processes. While early examples of fractals stem from the Earth sciences (such as the length of a coastline), the role of fractals in geology at present is mostly experimental in character. Care must be taken, however, in applying concepts as complex as fractals and chaos. Determining the quality of data sets and the suitability of a technique for the desired analysis can be treacherous. Simple scaling properties such as self-similarity and self-affinity and the results of more complex cascade scaling do not generally provide adequate descriptions of geologic surfaces, as in the case of the submarine topography example.

On the other hand, methods developed in the course of fractal analysis yield geologically meaningful results. Both scale-dependent variography and the double-trace moment technique provide ways to quantitatively describe scaling behavior and variability in spatial structure across scales. This should lead the way to the development of more complex mathematical models to capture the variability of geologic processes.

Acknowledgments

Thanks are extended to Steven Miller, University of California, Santa Barbara, for making the Deep Tow data available; to Daniel Lavallée, formerly at Meteorology Nationale, Paris, for the double-trace moment analysis program; and to Denelda Nollenberger and Jo Griffith of Scripps Institution of Oceanography for typing the manuscript and helping with illustrations. I gratefully acknowledge support provided by a Feodor Lynen Fellowship of the Alexander von Humboldt Foundation, Bonn, Germany.

References

Bak, P., 1990, Predicting earthquakes: *EOS Trans.*, Am. Geophys. Union, v. 71, p. 466.

Bell, T.H., 1975, Statistical features of seafloor topography: *Deep Sea Res.*, v. 22, p. 883-892.

Berkson, J.M. and Matthews, J.E., 1983, Statistical properties of seafloor roughness, *in* Pace, N.G., (ed.), *Acoustics in the Sea-bed*: Bath Univ. Press, Bath, England, p. 215-223.

Berry, M.V. and Lewis, Z.V., 1980, On the Weierstrass–Mandelbrot fractal function: *Proc. Royal Soc. London*, v. 370, Ser. A, p. 459-484.

Brown, S.R., Rundle, J.B. and Scholz, C.K., 1990, A simplified spring-block model of earthquakes: *EOS Trans.*, Am. Geophys. Union, v. 71, p. 467.

Brown, S.R. and Scholz, C.H., 1985, Broad band width study of the topography of natural rock surfaces: *Jour. Geophysical Research*, v. 90, p. 12575-12582.

Burrough, P.A., 1981, Fractal dimensions of landscapes and other environmental data: *Nature*, v. 294, no. 5838, p. 240-242.

Cantor, G., 1883, Grundlagen einer allgemeinen Mannichfaltigkeitslehre: *Math. Annalen*, v. 21, p. 545-591.

Caratheodory, C., 1914, Über das lineare Mass von Punktmengen—eine Verallgemeinerung des Längenbegriffs: *Nachr. der K. Gesellschaft der Wissenschaften zu Göttingen, Mathematisch-physikalische Klasse*, p. 404-426.

Chilès, J.P., 1988, Fractal and geostatistical method for modeling of a fracture network: *Math. Geol.*, v. 20, p. 631-654.

Culling, W.E.H., 1989, The characterization of regular and irregular surfaces in the soil-covered landscape by Gaussian random fields: *Computers & Geosci.*, v. 15, p. 219-226.

Elliot, J.K., 1989, An Investigation of the change in surface roughness through time on the foreland of Austre Okstindbreen, North Norway: *Computers & Geosci.*, v. 15, p. 209-217.

Fox, C.G., 1989, Empirically derived relationships between fractal dimension and power law from frequency spectra: *Pure Appl. Geophys.*, v. 131, p. 211–239.

Fox, C.G. and Hayes, D.E., 1985, Quantitative methods for analyzing the roughness of the seafloor: *Rev. Geophys. Space Phys.*, v. 23, p. 1–48.

Fricke, R. and Klein, F., 1897, *Vorlesungen Über die Theorie der Automorphen Functionen:* Teubner Verlag, Leipzig [2 volumes].

Ghil, M., 1990, Nonlinear oscillations, chaos and paleoclimate: *EOS Trans.*, Am. Geophys. Union, v. 71, p. 465.

Goff, J.A., 1991, A global and regional stochastic analysis of near-ridge abyssal hill morphology: *Jour. Geophysical Research*, Pt. B, v. 96, no. 13, p. 21,713–21,737.

Goff, J.A., Jordan, T.H., Edwards, M.H. and Fornari, D.J., 1991, Comparison of a stochastic seafloor model with SeaMARC II bathymetry and Sea Beam data near the East Pacific Rise 13° − 15° N: *Jour. Geophysical Research*, Pt. B, v. 96, no. 3, p. 3867–3885.

Goodchild, M.F., 1988, Lakes on fractal surfaces: A null hypothesis is for lake-rich landscapes: *Math. Geol.*, v. 20, p. 615-630.

Grassberger, P. and Procaccia, J., 1983, Measuring the strangeness of strange attractors: *Physica D.*, p. 189–208.

Hausdorff, F., 1919, Dimension und Äusseres Mass: *Math. Annalen*, v. 79, p. 157–179.

Herzfeld, U.C., 1992, Least squares collocation, geophysical inverse theory, and geostatistics: A bird's eye view: *Geophys. Jour. Internat.*, [in press].

Herzfeld, U.C., Kim, I.I. and Orcutt, J.A., 1992, Is the ocean floor a fractal?: *Jour. Geophysical Research*, [in press].

Holden, A.V., 1986, *Chaos:* Princeton Univ. Press, Princeton, New Jersey, 330 pp.

Hough, S.E., 1990, Estimating the fractal dimension of topographic profiles: *EOS Trans.*, Am. Geophys. Union, v. 71, p. 466.

Huang, J., 1990, Modeling seismic faulting as a chaotic dynamical system: *EOS Trans.*, Am. Geophys. Union, v. 71, p. 467.

Jones, J.G., Thomas, R.W. and Earwicker, P.G., 1989, Fractal properties of computer-generated and natural geophysical data: *Computers & Geosci.*, v. 15, p. 227–235.

Journel, A.G. and Huijbregts, C., 1989, *Mining Geostatistics:* Academic Press, London, 600 pp.

Julia, G., 1918, Memoire sur l'iteration des fonctions rationelles: *Math. Pures et Appliques*, v. 4, p. 47–245.

Kagan, Y.Y., 1991, Geometry of earthquake faulting: *EOS Trans.*, Am. Geophys. Union, v. 72, p. 58.

Krishnamurthy, V., 1990, Chaotic attractors with interannual variations in a climate model: *EOS Trans.*, Am. Geophys. Union, v. 71, p. 465.

Lavallée, D., Lovejoy, S., Ladoy, P. and Schertzer, D., 1991, Non-linear variability of landscape topography: Multifractal analysis and simulation, *in* De Cola, L. and Lam, N., (eds.), *Fractals in Geography:* Prentice Hall, New York, [in press].

Lebesque, H., 1903 [1972], Sur le probleme des aires, 1903: *Enseignment Mathématique*, v. 4, Genéve, p. 29–35.

Lovejoy, S. and Mandelbrot, B.B., 1985, Fractal properties of rain and a fractal model: *Tellus*, p. 209–232.

Lovejoy, S. and Schertzer, D., 1991, Multifractal analysis techniques and the rain and cloud fields from 10^{-3} to 10^6 m, *in* Schertzer, D. and Lovejoy, S., (eds.), *Nonlinear Variability, Scaling and Fractals:* Kluwer, Dordrecht, p. 111–144.

Luyendyk, B.P. and Macdonald, K.C., 1977, Physiography and structure of the inner floor of the FAMOUS rift valley: Observations with a deep-towed instrument package: *Geol. Soc. Amer. Bull.*, v. 88, p. 648–663.

Macdonald, K.C. and Luyendyk, B.P., 1977, Deep-tow studies of the structure of the Mid-Atlantic Ridge Crest near lat. 37° N: *Geol. Soc. Amer. Bull.*, v. 88, p. 621–636.

Malinverno, A., 1989, Segmentation of topographic profiles of the seafloor based on a self-affine model: *IEEE Trans.*, v. 14, p. 4.

Malinverno, A. and Gilbert, L.E., 1989, A stochastic model for the creation of abyssal hill topography at a slow spreading center: *Jour. Geophysical Research*, v. 94, p. 1665–1675.

Mandelbrot, B.B., 1964, Self-similar random processes and the range: *IBM Research Rept.*, v. RC-1163 [unpublished].

Mandelbrot, B.B., 1965, Self-similar error clusters in communication systems and the concept of conditional stationarity: *IEEE Trans. Communications Technology*, v. 13, p. 71–90.

Mandelbrot, B.B., 1967, How long is the coast of Britain? Statistical self-similarity and fractional dimension: *Science*, v. 156, p. 636–638.

Mandelbrot, B.B., 1974, Intermittent turbulence in self-similar cascades: Divergence of high moments and dimension of the carrier: *Jour. Fluid Tech.*, v. 62, p. 331–358.

Mandelbrot, B.B., 1975, *Les Objets Fractals: Forme, Hasard et Dimension:* Flammarion, Paris [republished 1989], 286 pp.

Mandelbrot, B.B., 1983, *The Fractal Geometry of Nature:* W.H. Freeman and Co., New York, 460 pp.

Matheron, G., 1963, Principles of Geostatistics: *Econ. Geol.*, v. 58, p. 1246–1266.

Matteucci, G., 1990, A study of the climatic variability of the full Pleistocene using a stochastic resonance model: *EOS Trans.*, Am. Geophys. Union, v. 71, p. 465.

Mundt, M.D., Maguire, W.B. and Chase, R.R.P., 1990, Chaos in the sunspot cycle: Analysis and Prediction: *Jour. Geophysical Research (Space Phys.)*, [in press].

Mutter, J.C., Morris, E. and Dettrick, R.S., 1990, Structures imaged in the North Atlantic oceanic crust exhibit self-similarity—Why?: *EOS Trans.*, Am. Geophys. Union, v. 71, p. 615.

Nicholl, M., 1990, Is Old Faithful a strange attractor?: *EOS Trans.*, Am. Geophys. Union, v. 71, p. 466.

Nicolis, C. and Nicolis, G., 1985, Gibt es einen Klima-Attraktor: *Phys. Bl.*, v. 41, p. 5–9.

Olea, R.A., (ed.), 1991, *Geostatistical Glossary and Multilingual Dictionary:* Studies in Mathematical Geology No. 3, Oxford Univ. Press, New York, p. 177.

Peano, G., 1890, Sur une courbe, qui remplit une aire plane: *Math. Annalen*, v. 36, p. 157–160.

Peitgen, H.O. and Richter, P.H., 1985, Fractale Strukturen, Mandelbrot-Menge und die Theorie der Phasenübergänge: *Phys. Bl.*, v. 41, p. 19–20.

Peitgen, H.O. and Richter, P.H., (eds.), 1986, *The Beauty of Fractals:* Springer–Verlag, Berlin, 199 pp.

Peitgen, H.O. and Saupe, D., (eds.), 1988, *The Science of Fractal Images:* Springer–Verlag, Berlin, 325 pp.

Sayles, R.S. and Thomas, T.R., 1978, Surface topography as a non-stationary random process: *Nature*, v. 271, p. 431–434.

Smith, D.K. and Shaw, P.R., 1990, Seafloor topography: A record of a chaotic dynamical system?: *Geophys. Res. Lett.*, v. 17, p. 1541–1544.

Spiess, F.N. and Lonsdale, P., 1982, Deep Tow rise crest exploration techniques: *Mar. Technol. Soc. Jour.*, v. 16, p. 67–74.

Steacy, S.J. and Sammis, C.G., 1990, Damage mechanics of a fractal fault zone: *EOS Trans.*, Am. Geophys. Union, v. 71, p. 467.

Stewart, C., 1990, The route to chaotic mantle convection: *EOS Trans.*, Am. Geophys. Union, v. 71, p. 466.

Tessier, Y., Lovejoy, S. and Schertzer, D., 1991, Universal multifractals: Theory and observations for rain and clouds: *Jour. Appl. Meteorol.*, [in press].

Tsonis, A.A. and Elsner, J.B., 1990, Multiple attractors, fractal basins and long-term climate dynamics: *Beiträge zur Physik der Atmosphäre*, [in press].

Turcotte, D.L., 1986, A fractal model for crustal deformation: *Tectonophysics*, v. 132, p. 261–269.

Unwin, D., 1989, Fractals and the geosciences. Introduction: *Computers & Geosci.*, v. 15, p. 163–165.

Varnes, D.J., 1991, The cyclic and fractal seismic series preceding the February 14, 1980, M4.8 earthquake near the Virgin Islands: *EOS Trans.*, Am. Geophys. Union, v. 72, p. 58.

Wong, P. and Lin, J., 1988, Studying fractal geometry on submicron length scales by small-angle scattering: *Math. Geol.*, v. 20, p. 655-665.

Yfantis, E.A., Flatman, G.T. and Englund, E.J., 1988, Simulation of geological surfaces using fractals: *Math. Geol.*, v. 20, p. 667–672.

19

AN EXECUTABLE NOTATION, WITH ILLUSTRATIONS FROM ELEMENTARY CRYSTALLOGRAPHY

Donald B. McIntyre

Elementary crystallography is an ideal context for introducing students to mathematical geology. Students meet crystallography early because rocks are made of crystalline minerals. Moreover, morphological crystallography is largely the study of lines and planes in real three-dimensional space, and visualizing the relationships is excellent training for other aspects of geology; many algorithms learned in crystallography (*e.g.*, rotation of arrays) apply also to structural geology and plate tectonics.

Sets of lines and planes should be treated as entities, and crystallography is an ideal environment for introducing what Sylvester (1884) called "Universal Algebra or the Algebra of multiple quantity." In modern terminology, we need SIMD (Single Instruction, Multiple Data) or even MIMD. This approach, initiated by W.H. Bond in 1946, dispels the mysticism unnecessarily associated with Miller indices and the reciprocal lattice; edges and face-normals are vectors in the same space.

A Simple, Consistent, Executable Notation

The growth of mathematical notation has been haphazard, new symbols often being introduced before the full significance of the functions they represent had been understood (Cajori, 1951; McIntyre, 1991*b*). Iverson introduced a consistent notation in 1960 (*e.g.*, Iverson 1960, 1962, 1980). His language, greatly extended in the executable form called *J* (Iverson, 1993), is used here. For information on its availability as shareware, see the Appendix. Publications suitable as tutorials in *J* are available (*e.g.*, Iverson, 1991; McIntyre, 1991*a, b*; 1992*a, b, c*; 1993).

Crystallographic Axes

Crystals are periodic structures consisting of unit cells (parallelepipeds) repeated by translation along axes parallel to the cell edges. These edges define the crystallographic axes. In a crystal of cubic symmetry they are orthogonal and equal in length (Cartesian). Those of a triclinic crystal, on the other hand, are unequal in length and not at right angles. The triclinic system is the general case; others are special cases.

The formal description of a crystal gives prominent place to the lengths of the axes (a, b, and c) and the interaxial angles (α, β, and γ). A *canonical form* groups these values into a 2×3 table (matrix), the first row being the lengths and the second the angles. For example, the canonical form of chalcanthite, $CuSO_4 \cdot 5H_2O$ (Dana, 1951, Vol. 2, p. 489), is:

```
<"0 chalcanthite=. 6.11 10.673 5.95,: 97.583 107.167 77.55
```

6.11	10.673	5.95
97.583	107.167	77.55

```
(;:'a b c'),: ;:'alpha beta gamma'
```

a	b	c
alpha	beta	gamma

The corners of the unit cell are lattice points, *i.e.*, points with identical environments. Further points are found by taking integer multiples of each axis and adding these in all combinations. The resulting array is the *direct lattice*. The axes can be referred to an imaginary Cartesian framework, x, y, and z, with its origin at a corner of the unit cell. By convention the $+x$-axis is towards the observer, the $+y$-axis to the right, and the $+z$-axis vertically upwards. In the two-dimensional example of Figure 1, the coordinates of the lattice points are given with reference both to Cartesian reference axes, x and y, and to crystallographic axes, a and b.

The Cartesian coordinates of the crystallographic axes a and b are:
```
a=. 1.75 0.5
b=. 0.25 2
```

The lattice is therefore defined by a matrix; the items (rows) refer to the axes and the columns give the Cartesian coordinates. I call it the "d-matrix," because it describes the *direct lattice*, as opposed to the *reciprocal lattice* of X-ray crystallographers. Bond was first to point out the computational advantage of matrix methods in crystallography (Bond, 1946; Terpstra and Codd, 1961; McIntyre, 1978).

```
<"0 d=. a,:b
```

1.75	0.5
0.25	2

Points on the lattice rows that extend the crystallographic axes are given by integer multiples of a and b:

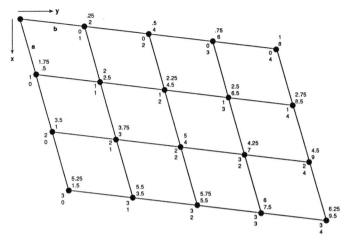

Figure 1: Two-dimensional example showing lattice points with both Cartesian reference axes, x and y (coordinates on upper right), and crystallographic axes, a and b (coordinates on lower left).

```
       ]p=. a*/i.4                    ]q=. b*/i.5
 0 1.75 3.5 5.25             0 0.25 0.5 0.75 1
 0  0.5   1 1.5              0   2    4    6 8
```

The *outer product* of p and q gives other lattice points:

```
       $p+/q
 2 4 2 5
```

The shape of the outer product is 2 4 2 5 because $p + /q$ not only adds x to x and y to y, but x to y and y to x. A dyadic transpose takes a rank-3 cut through this rank-4 array to give the Cartesian coordinates of the lattice points:

```
       $x=.  (<0 2)|:(a*/i.4)+/(b*/i.5)
 4 5 2

         x
    0    0
 0.25    2
 0.5     4
 0.75    6
    1    8
```

etc.

The coordinates can also be computed using an *inner product*:

```
ip=. +/ .*
d ip~ 4 5#:i.20
```

Vector Lengths and Angles

Because the norm (length) of a vector is the square root of the sum of squares of its components, the axial lengths are computed easily from the d-matrix:

```
norm=. +/&.*:"1
norm d
1.82003 2.01556
```

Dividing each axis by its norm, we obtain the *direction vectors*:

```
dv=. %"1 0 norm
dv d
0.961524 0.274721
0.124035 0.992278
```

Define the function *angles* to compute interaxial angles:

```
ipt=. ip |:         NB. inner product with transpose
arcos=. _2&o.       NB. arcosine:  angle in radians
real=. {."1@+.      NB. imaginary parts may come from rounding
rfd=. %&180@o.      NB. radians from degrees
dfr=. rfd^:_1       NB. degrees from radians (inverse of rfd)
clean=. ] * (<:|)   NB. set small values to zero
angles=. 1e_6&clean@dfr@real@arcos@ipt@dv
angles d
       0 66.9296
66.9296        0
```

The result gives the angles between each item (row) in the argument and every other item, though in this case there are only two items. The angle between the crystallographic *a*-axis and the Cartesian *x*-axis is given by:

```
dfr arcos 1 0 ip dv a
15.9454
```

Rotation to a Standard Setting

The lattice is rotated with respect to the Cartesian axes so that the *a*-axis is parallel to the *x*-axis (or lies in the *x*–*z* plane in three dimensions). We achieve this with a rotation matrix of the form:

cos	-sin
sin	cos

```
rot=. '(cos,-sin),:(''sin'';''cos'')'=. 1 2 o. rfd y.' : ''
```

or alternatively:

```
rot=. (1 _1,:1 1)&*@((2 1,:1 2)&o. @ rfd)
```

A table of old and new coordinates is given by:

```
6.2 6.2 8.2 6.2 ": x,"1 x ip rot 15.9454
0.00  0.00   0.00  0.00
0.25  2.00   0.79  1.85
0.50  4.00   1.58  3.71
    etc.
```

In the standard setting of a triclinic crystal the *c*-axis is oriented parallel to the vertical *z*-axis; the crystal is then rotated about *z* until the *a*-axis lies in the *x*–*z* plane with +*a* towards the observer and the angle between +*a* and +*z* not less than 90°. This fixes the position of the *b*-axis.

Computing the d-Matrix from the Canonical Form

Every crystal description should include the d-matrix, but as this is rarely done; the function *dmat* is given here so that d-matrices can be computed from canonical forms. Terpstra and Codd (1961) give the necessary spherical trigonometry.

```
rfd=. %&180@o.          NB. radians from degrees
sin=. 1&o.              NB. sine of angle in radians
cos=. 2&o.              NB. cosine of angle in radians
SinCos=. 1 2&o.         NB. sine and cosine
```

Crystallographic axes (lengths) or interaxial angles are:

```
a=. 0&{
b=. 1&{
c=. 2&{
ab=. a,b
axisa=. 1 0 2&{@(@(0&,@SinCos@b)
```

and cos(rho) and cos(sigma) are based on Terpstra and Codd (1961, p. 287)

```
CosRho=. (cos@c - */@cos@ab) % */@sin@b
CosSigma=. sin@b %~ %:@(>:@+:@(*/@cos) - +/@*:@cos)
axisb=. CosRho,CosSigma,cos@a

dm=. ,&0 0 1 @(axisa,:axisb)@rfd@b
dmat=. ({. *"0 1 dm) f.
```

Test data, chalcanthite from J.D. Dana (1951, Vol. 2, p. 489), ch (chalcanthite) and anorthite from Terpstra and Codd (1961, p. 290), are:

```
chalcanthite=. 6.11 10.673 5.95,: 97.583 107.167 77.55
ch=. 0.5705 1 0.5565,: 82.367 107.433 102.55
anorthite=. 0.6344 1 0.5505,:93.15 115.9833 91.2
```

Note that *dmat* produces the d-matrix for any number of crystals simultaneously, *e.g.*,

```
   dmat chalcanthite
5.83779       0 _1.80341
1.97316 10.394 _1.40843
      0      0    5.95

   dmat"2 chalcanthite, ch,: anorthite
```

This expression computes the d-matrices for each crystal.

Computing the Canonical Form from the d-Matrix

```
canon=. norm,:5 2 1&{@(@,@angles
```

This is the inverse of *dmat*:

```
   chalcanthite-: canon dmat chalcanthite
1
```

Volume of the Unit Cell

The determinant of the d-matrix is the two-dimensional area or three-dimensional volume of the unit cell:

```
    vol=. det=. -/ .*
    vol dmat chalcanthite
361.035
```

Cozonal Faces, and the Angles on a Cubic Crystal

It follows that the determinant can be used to discover whether three crystal faces are cozonal, *i.e.*, whether their normals are coplanar:

```
    det 1 2 2, 3 1 1,: 0 1 1
  0
    det 1 2 1, 3 1 1,: 0 1 1
 _3
```

The faces in the first set are cozonal; those in the second are not. Because the axes of a cubic crystal are already Cartesian, the function (verb) *angles* determines the angles between face normals:

```
    cubic=. 1 0 0,1 1 0,1 1 1,0 1 0,0 1 1,1,1 _1 1,: 3 1 2
    (3 ":cubic),"( 1) 8.2 ": 1e_4 clean angles cubic
1 0 0    0.00    45.00    54.74    90.00    90.00    54.74    36.70
1 1 0    45.00    0.00    35.26    45.00    60.00    90.00    40.89
1 1 1    54.74    35.26    0.00    54.74    35.26    70.53    22.21
    etc.
```

The angles from the front face of the cube to the other faces are given in degrees, minutes, and seconds by:

```
4 ": cubic,"(1) 0 60 60 #: 60*60* 0{angles cubic
1    0 0    0    0    0
1    1 0   45    0    0
1    1 1   54   44    8
    etc.
```

The Direct Lattice: Bond Lengths and Angles

The positions of atoms within the cell are given with respect to the axes of the direct lattice; for example (0 0 0) is a corner and (0.5 0.5 0.5) the center of the cell, irrespective of cell size or shape. Because edges have simple rational indices with respect to the direct lattice, calculations involving either edges or atomic positions are straightforward. Consider the tetragonal mineral rutile (TiO_2) (Bunn, 1961, p. 226) with d-matrix:

```
d=. 4.58 0 0,0 4.58 0,:0 0 2.98
```

A titanium at (0.5 0.5 0.5) is surrounded by six oxygens, with coordinates:

```
O=. 0.31 0.31 0, 0.69 0.69 0,: 0.81 0.19 0.5
O=. 0, 0.19 0.81 0.5, 0.31 0.31 1,: 0.69 0.69 1
```

To find the Ti-O bond lengths, move the origin to the central titanium, and convert to Cartesian coordinates.

```
6.2": 0,"1 0 norm x=. (0-0.5) ip d
0.31  0.31  0.00 1.93
0.69  0.69  0.00 1.93
0.81  0.19  0.50 2.01
```
   etc.

Angles between bonds radiating from the titanium are given by:

```
BondAngles=. 9.2&":@angles@dv
BondAngles x
    0.00   79.11  90.00 90.00   100.89   180.00
```
   etc.

The Reciprocal Lattice

The axes of the reciprocal lattice are normal to the faces of the unit cell and their lengths are the reciprocals of the spacings of planes in the set, *i.e.*, the distance between a face on one side of the unit cell and the corresponding face on the other side. Because they determine whether X-rays of given wavelength will be reflected, these distances (*d-spacings*) are important in X-ray crystallography.

In the direct lattice, edges are vectors whose coordinates are simple rational numbers, whereas in the reciprocal lattice face-normals are vectors whose coordinates are the Miller indices. Consequently problems formerly requiring considerable ingenuity and skill in spherical trigonometry can now be solved by simple matrix methods.

The d-matrix of chalcanthite (data from Berry and Mason, 1959, p. 435) is:

```
d=. 5.84927 0 _1.80693, 1.97722 10.4155 _1.41139,: 0 0 5.96
```

The reciprocal lattice is defined by a corresponding matrix (the *r-matrix*) which gives the coordinates of the axes (a^*, b^*, c^*) of the reciprocal lattice with respect to the Cartesian frame. The r-matrix is simply the transpose of the inverse of the d-matrix, a relationship worth exploring, but space does not allow further elaboration here.

For chalcanthite (Berry and Mason's data), we have:

```
r=. 1e_10 clean |: %. d
```

	x	y	z
a^*	0.170962	_0.0324545	0
b^*	0	0.0960112	0
c^*	0.0518314	0.0128971	0.167785

Because the axes are expressed with reference to a Cartesian frame, their lengths and interaxial angles are obtained as they were for the d-matrix:

```
canon r
0.174015 0.0960108 0.176082
85.7997   74.0061   100.749
```

Interfacial Angles

Berry and Mason give some of the angles between the following faces:

```
m=. 1 0 0, 1 _1 0, _1 _1 1, 0 1 0, 1 1 0,: 1 3 0
```

Because the inner product of Miller indices and the r-matrix converts the face-normals to a Cartesian frame, all these angles are easily computed:

```
     (3 3 3,6#9.2)":m,"(1) angles m ip r
 1  0  0      0.00 26.17   120.51   100.75    31.14    66.97
     etc.
```

Computation of d-Spacing in a Triclinic Crystal

The high symmetry of cubic crystals makes computation of the spacings of crystal planes easy, but calculations become more difficult with decreasing symmetry. As Bunn (1945, p. 378; 1961, p. 456) pointed out: "For monoclinic and triclinic cells, the formulae for the spacings are very unwieldy. Graphical methods based on the conception of the reciprocal lattice are recommended."

This difficulty vanishes when we follow Bond in recognizing Miller indices as defining true vectors in the reciprocal lattice. The Miller indices are converted to Cartesian by taking the inner product with the r-matrix, and calculation of lengths and angles is then trivial. As an example, consider chalcanthite. The Joint Committee on X-ray Powder Diffraction Standards uses a different cell (JCPDS 11–646) from that given by Dana. Computing the r-matrix from the JCPDS data:

```
r=. : %. dmat 7.155 10.71 5.955,: 97.63 125.32 94.32
```

JCPDS give the first 11 lines of the diffraction pattern as:

```
hkl=. 0 1 0,1 0 0,1 0 _1,1 _1 0,0 2 0,1 1 _1,1 _1 _1,0 1 _1,
      0 0 1,1 1 0,:1 _2 0
```

The d-spacings of these planes are then:

```
     3 3 3 8.2 ": hkl,"1 0 % norm hkl ip r
 0  1  0    10.43
 1  0  0     5.73
 1  0 _1     5.67
     etc.
```

Transformation of Settings

When crystals have low symmetry (monoclinic and triclinic), authorities often differ in their choice of unit cell. The determinants of the various d-matrices make it possible to compare the cell volumes. Transformation matrices make it simple to convert Miller indices from one setting to another. Moreover, if a face has the same indices in two settings, its normal must be an eigenvector of the transformation matrix. Space does not allow examples here.

Conclusions

Despite the advantages of Bond's matrix methods, textbooks continue to give only the canonical form: axial lengths and interaxial angles. This is unfortunate, because it is easier to go from the d-matrix to canonical form than to go in the opposite direction. Starting with the d-matrix, we can use matrix methods immediately; whereas if we begin with the canonical form, we must use algorithms derived from spherical trigonometry.

In the review of a mineralogy textbook, Derek Flinn (1972) wrote:

> "I believe that the mystical approach to Miller indices should be abandoned and that the student should be told that they are face-normal vectors, that they belong to the same vector space as the zone axis indices and that the addition and multiplication of these indices are elementary operations in vector algebra."

Quoting this 15 years ago, I said, "Flinn is correct, and I can testify that APL makes what he recommends easy and natural" (McIntyre, 1978, p. 250). Today *J* provides further testimony.

In his 1977 Turing Award paper, John Backus (1978) advocated a new functional style of programming. I have shown here how this style has been achieved in a consistent, executable notation in which there is no explicit reference to function arguments (Hui *et al.*, 1991).

Acknowledgments

Kenneth Iverson, Roger Hui, and E.E. McDonnell went out of their way to tutor me in *J*, but they are not responsible for any stylistic failures in the use of their language.

I am indebted to many students who took my crystallography class at Pomona College, California, and in particular to George Clark '46, Donald McIsaac '59, James Kelley '63, Steve Norwick '65, David Pollard '65, Jim Kauahikaua '73, Paul Delaney '73, Allen Glazner '76, Tom Hoisch '79, and Peter Christiansen '87, for their special interest and stimulation. My colleague Paul Yale, Mathematics Department, gave suggestions and encouragement over many years. Through the support of the Mellon Foundation, Don L. Orth, IBM, audited my class at a critical time 17 years ago when we had newly adopted Iverson's system of Direct Definition, then implemented on one of IBM's earliest desktop computers.

References

Backus, John, 1978, Can programming be liberated from the Von Neumann style? A functional style and its algebra of programs [1977 Turing Award Paper]: *Comm. of the A.C.M.*, v. 21, no. 8, p. 613–641.

Berry, L.G. and Mason, Brian, 1959, *Mineralogy:* W.H. Freeman and Co., San Francisco, 612 pp.

Bond, W.L., 1946, Computation of interfacial angles, interzonal angles, and clinographic projection by matrix methods: *American Mineralogist*, v. 31, p. 31–42.

Bunn, C.W., 1945, *Chemical Crystallography:* Oxford University Press, New York [2nd Edition (1961), Appendix 2, "The spacings of crystal planes"], 422 pp.

Cajori, Florian, 1951, *A History of Mathematical Notations, Vol. 1:* [First published 1928.] The Open Court Publishing Co., La Salle, Illinois, 451 pp.

Cajori, Florian, 1952, *A History of Mathematical Notations, Vol. 2:* [First published 1929.] The Open Court Publishing Co., La Salle, Illinois, 367 pp.

Dana, J.D., 1951, *The System of Mineralogy, Vol. 2:* [7th Ed.] John Wiley & Sons, New York, 1124 pp.

Flinn, Derek, 1972, Review of M.H. Battey's "Mineralogy for Students": *Geological Journal*, v. 8 (April), p. *v–vi*.

Hui, R.K.W., Iverson, K.E. and McDonnell, E.E., 1991, Tacit Definition: *APL Quote Quad*, v. 21, no. 4, Assoc. Computing Machinery Press, p. 202–211.

Iverson, K.E., 1960, The description of finite sequential processes, *in* Cherry, Colin and Jackson, Willis, (eds.), *Proceedings of a Conference on Information Theory:* Imperial College, London, August, 1960, p. 447–457.

Iverson, K.E., 1962, *A Programming Language:* John Wiley & Sons, New York, 286 pp.

Iverson, K.E., 1980, Notation as a tool of thought [1979 Turing Award Paper]: *Comm. of the A.C.M.*, v. 23, no. 8, p. 444–465.

Iverson, K.E., 1991, *Programming in J:* Iverson Software Inc., Toronto, 72 pp.

Iverson, K.E., 1993, *J: Introduction and Dictionary, Version 6.2:* Iverson Software Inc., Toronto, 105 pp.

McIntyre, D.B., 1978, *The Architectural Elegance of Crystals Made Clear by APL:* Proceedings of an APL Users' Meeting, Toronto, September, 1978, I.P. Sharp Associates Ltd., p. 233–250.

McIntyre, D.B., 1991*a*, Mastering *J: APL Quote Quad*, v. 21, no. 4, Assoc. Computing Machinery Press, p. 264-273.

McIntyre, D.B., 1991*b*, Language as an intellectual tool: From hieroglyphics to APL: *IBM Systems Journal*, v. 30, no. 4, p. 554–581.

McIntyre, D.B., 1992*a*, Hooks and forks and the teaching of elementary arithmetic: *Vector*, v. 8, no. 3, p. 101–123.

McIntyre, D.B., 1992*b*, Using *J* with external data: Two Examples: *Vector*, v. 8, no. 4, p. 97–110.

McIntyre, D.B., 1992*c*, Using *J*'s boxed arrays: *Vector*, v. 9, no. 1, p. 92–105.

McIntyre, D.B., 1993, Jacobi's method for eigenvalues: an Illustration of *J*, [in press].

Terpstra, P. and Codd, L.W., 1961, *Crystallometry:* Academic Press, New York, 420 pp.

Appendix

For a summary of *J* notation see Hui *et al.* (1991). For a complete description see Iverson (1993). Words used as *J* symbols are spelled as single ASCII characters either alone or immediately followed by a period or colon. *J* is available as shareware for most computers from Iverson Software Inc., 33 Major Street, Toronto, Ontario M5S 2K9, Canada. Examples given in this paper were revised to execute with Version 6.1 (Nov. 1992). The *J* system for IBM-compatible machines and script files of tutorial material including executable versions of all examples in this paper are available on the *J Conference* maintained by ALMAC BBS Ltd., 141 Bo'ness Road, Grangemouth FK3 9BF, Scotland. To access the *J Conference*, telephone +44 (0) 324-665-371 and enter J 398 after completing the two online questionnaires.

20

UNCERTAINTY IN GEOLOGY

C. John Mann

The nuclear waste programs of the United States and other countries have forced geologists to think specifically about probabilities of natural events, because the legal requirements to license repositories mandate a probabilistic standard (US EPA, 1985). In addition, uncertainties associated with these probabilities and the predicted performance of a geologic repository must be stated clearly in quantitative terms, as far as possible.

Geoscientists rarely have thought in terms of stochasticity or clearly stated uncertainties for their results. All scientists are taught to acknowledge uncertainty and to specify the quantitative uncertainty in each derived or measured value, but this has seldom been done in geology. Thus, the nuclear waste disposal program is forcing us to do now what we should have been doing all along: acknowledge in quantitative terms what uncertainty is associated with each quantity that is employed, whether deterministically or probabilistically.

Types of Uncertainty

Uncertainty is a simple concept ostensibly understood to mean that which is indeterminate, not certain, containing doubt, indefinite, problematical, not reliable, or dubious. However, uncertainty in a scientific sense demonstrates a complexity which often is unappreciated. Some types of uncertainty are difficult to handle, if they must be quantified, and a completely satisfactory treatment may be impossible.

Initially, only uncertainty associated with measurement was quantified. The Gaussian, or normal, probability density function (pdf) was recognized by Carl Friedrich Gauss as he studied errors in his measurements two centuries ago and developed a theory of errors still being used today. This was the only type of uncertainty that scientists acknowledged until Heisenberg stated his famous uncertainty principle in 1928. As information theory evolved during and after World War II, major advances were made in semantic uncertainty. Today, two major types of uncertainty are generally recognized (Klir and Folger, 1988): ambiguity or nonspecificity and vagueness or fuzziness. These can be subdivided further into seven types having various measures of uncertainty based on probability theory, set theory, fuzzy-set theory, and possibility theory.

Most workers have associated, and confused, types of uncertainty with sources of uncertainty. Commonly, uncertainties that arise from parameters, data, models, computations, and human errors are recognized. Major disadvantages of this practice are: (1) Sources of uncertainty are numerous and therefore the types are nearly endless. This does not lead to simplification or general categorization. (2) Methods of handling various types of uncertainty are neither clear nor easy.

Cox (1982) considered three sources of uncertainty: artifactual, imperfect observations, and actual variance. Artifactual uncertainties are due to the analysis; imperfect observations are those that are either incomplete or unspecific; and actual variance is due to a parameter having a pdf rather than a single, constant value (Cox, 1982). A workshop sponsored by the Nuclear Energy Agency (OECD, 1987, p. 12) recognized two types of uncertainty, one type arising from random variability and a second type due to "lack of knowledge." Here, a new classification of types of uncertainty is followed, similar to that of Cox. Three types of uncertainty are recognized because each type arises in a different manner and is treated differently.

Type I—Type 1 uncertainty arises through error, bias, and imprecision of measurement processes. All processes of measurement will result in some, perhaps small, bias and imprecision (Eisenhart, 1968), so every value should have an associated plus and minus error. Some values are known better than others depending upon measurement devices, techniques, calibrations, and repetitions. However, all values could be known more accurately if better instrumentation were available or more time and effort were spent to reduce errors and improve calibration between different laboratories and methodologies. Methods for treating uncertainty due to errors are well developed; several excellent texts discuss these methods rigorously (Beers, 1957; Barford, 1985; Jaech, 1985), as do many papers (*e.g.*, Eisenhart, 1963).

Type II—A second type of uncertainty (Type II) arises from inherent variation in natural parameters. Uncertainty due to stochasticity and randomness initially was thought to be an artifact of the inability to measure with sufficient accuracy. Heisenberg's uncertainty principle was recognized by physicists (Heisenberg, 1930) but was thought to be true only at small scales. Recently the importance of Type II uncertainty has begun to be appreciated by geologists and biologists. Many natural processes, events, and phenomena now are recognized as inherently stochastic.

In geology, stochastic uncertainty arises from inhomogeneity and anisotropy in rock units—different physical properties exist in different spatial portions. If these variations could be determined accurately (normally they cannot) presumably they could be modeled deterministically. Conversely, probability density functions (pdf's) required to model stochastic variations are not sufficiently well determined so they can be used. Hence, additional uncertainty is introduced because of an inability to model accurately. Either average values or "effective" values are used in deterministic modeling, or poorly determined pdf's are used in stochastic modeling.

Natural variation at scales larger than that for which Heisenberg's principle is valid might become known if sufficient measurements were taken, but this is not certain and, in some cases, is definitely incorrect. If an attempt is made to determine pdf's adequately, Heisenberg's principle is seen to be valid on a geological megascale because many variables cannot be measured without altering the rock. However, stochasticity usually can be handled adequately by stochastic modeling. Stochastic hydrology, for example, provides ground-water predictions that are as accurate as deterministic computations based on considerably more field data (Gutjahr, 1992).

Type III—Another type of uncertainty is due to lack of knowledge, or scientific ignorance (Type III). Even if all measured values were known exactly, if stochasticity was not present, and if deterministic equations were used for all calculations, some uncertainty would remain because relationships among all components are not known or known incorrectly. Uncertainty may arise in calculations because the analog used in numerical analysis is imperfect or the conceptual mathematical model is incorrect or imperfect. Thus, Type III uncertainty arises from a) limited knowledge and b) inability to model nature accurately. Type III uncertainty is present when predictions of the future are made because all necessary parameters will never be known nor estimated accurately. Geological deductions about the past also have associated uncertainties because past conditions and parameters can never be known exactly. The uncertainty is proportional to the length of time into the future or past over which predictions are made.

Type III uncertainty has been ignored almost entirely because it is theoretically unknowable. This is true also for bias in Type I uncertainty if independent means of measurement are not available. Nonetheless, bias must be estimated (Eisenhart, 1963) and geologists are now forced to evaluate Type III uncertainty for risk assessments of nuclear waste repositories. These are subjective probability estimates made by experts who are expressing their confidence in current knowledge, understanding of natural processes, and techniques available for analysis and interpretation (Apostolakis, 1990).

Table 1: Sources of Uncertainty

Type I Uncertainty (errors, bias, imprecision)

Errors in measurement (gross, bias, operator, conceptual, *etc.*)	Inadequate sampling
	Physical limitation to sampling
Conceptual error in population measured	Inability to determine accurate pdf's
Bias in measurement process (device, method, technique)	Inability to know true accuracy (true bias)
	Inability to truly isolate a system
Imprecision of measurement process (device, method, technique)	Computational inaccuracies (mathematical analog imperfect)

Type II Uncertainty (stochasticity)

Inherent natural variation	Physical inability to sample adequately
Heterogeneity in materials	Practical need to use average values or other indicators rather than pdf's
Anisotropy in parameters	
Inability to characterize a variable adequately	Noise in natural system
	Noise in computational system
Inability to determine an accurate pdf for a variable	

Type III Uncertainty (ignorance)

A. Lack of knowledge

Incomplete knowledge	Dissonance in evidence
Erroneous knowledge	Nonspecificity in evidence
Imperfect concepts, laws, hypotheses, and principles	
	B. Need for generalizations
Use of subjective probabilities rather than objective probabilities	Need for simplifications
	Use of incorrect models (conceptual or mathematical)
Ambiguity in concepts, data, models	
Vagueness in concepts, data, models	Use of imperfect models (conceptual or mathematical)
Fuzziness in concepts, data, models	
Distortion in concepts, data, models	Computational inaccuracies (mathematical analog imperfect)
Confusion in evidence	

Sources of Uncertainty

Innumerable sources of uncertainty can be identified in a detailed analysis of any specific system or geologic process, event, or phenomenon. However, sources cannot be measured individually and it is sufficient to compound them all into an accurate estimate of total uncertainty. A simplified list of common and generally present sources of uncertainty (Table 1) reveals that a single source may contribute to different types of uncertainty (*e.g.*, sampling, conceptual model, and errors). Hence, the three types of uncertainty are not mutually exclusive or statistically independent.

Handling and Propagating Uncertainty

Most familiar to geologists are Type I and Type II uncertainty which are both quantitative, as opposed to Type III which is qualitative. Uncertainty due to measurement processes can be minimized by careful and repeated measurements, calibration of equipment and methodology, determination of analytical imprecision, and comparison with other, independent measurement methods (Eisenhart, 1963; Laslett and Sandland, 1989). If bias or inaccuracy cannot be determined in some objective manner, it must be estimated in a subjective fashion.

Type I uncertainty can be propagated deterministically to a result, r, for

$$r = f(X_1, X_2, X_3, \ldots, X_j)$$

by

$$U_r = \left[\left(\frac{\partial r}{\partial X_1} U_{X_1} \right)^2 + \left(\frac{\partial r}{\partial X_2} U_{X_2} \right)^2 + \cdots \left(\frac{\partial r}{\partial X_j} U_{X_j} \right)^2 \right]^{1/2}$$

where U_{X_i} is uncertainty in x_i and U_r is uncertainty in r (Coleman and Steele, 1989). This simplifies for

$$r = kX_1^a, X_2^b, X_3^c, \ldots$$

to

$$\left(\frac{U_r}{r} \right)^2 = a^2 \left(\frac{U_{X_1}}{X_1} \right)^2 + b^2 \left(\frac{U_{X_2}}{X_2} \right)^2 + c^2 \left(\frac{U_{X_3}}{X_3} \right)^2 + \cdots \quad .$$

Type II, or stochastic, uncertainty may be more difficult to quantify because it can be measured directly only for geologic phenomena which occur rapidly (say, over less than 200 years) and are measured repeatedly. For slow processes and those which occurred in the past, stochasticity can be inferred only indirectly. For example, results from stochastic models can be compared with empirical data and adjusted by suitable model changes until a satisfactory agreement is reached. Monte Carlo simulations will establish what variation can be expected to occur. If the model is a reasonable representation of the natural system, the Monte Carlo variations will be realistic. The accuracy of the model can be determined only by repeated comparisons with the natural system under different conditions, which may not be possible.

Both Type I and II uncertainty can be described completely by probability density functions for errors, precision, and stochasticity and are measured and propagated in a probabilistic framework. Errors and precision ordinarily approximate a Gaussian distribution; however, this should not be assumed uncritically because variables which are not distributed symmetrically may have asymmetric error distributions, especially if bounded on one extremity. Geological data characteristically are asymmetrical and may span several orders of magnitude. Natural stochasticity typically has a non-Gaussian distribution, often log normal, binomial, log-hyperbolic, gamma, or various forms of a beta distribution (Fig. 1).

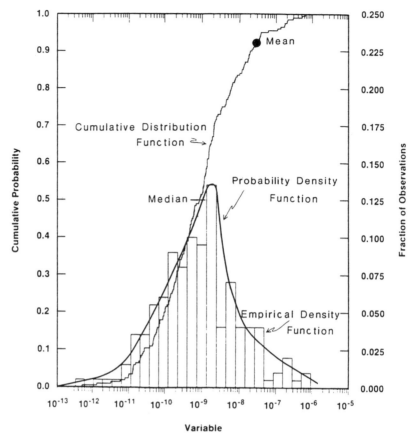

Figure 1: A typical geologic variable as illustrated by its 1) edf, may approach a log normal distribution spanning several orders of magnitude; 2) cdf; and 3) an asymmetric pdf which may or may not be approximated by a well-known theoretical curve.

Bias, or accuracy, of Type I uncertainty is a constant correction factor which must be applied to the result of a specific measurement process. The correction is not necessarily the same in either quantity or sign for supposedly identical instruments.

Type III uncertainty is called "qualitative uncertainty" because it was first studied in connection with information and communication and is associated closely with semantic ambiguity and fuzziness. Most workers in information theory do not think this type of uncertainty can be handled probabilistically, but must be treated by possibility theory or fuzzy-set theory (Smithson, 1989). Consequently, treatment of uncertainty due to ignorance is much different than the characterization of Type I and II uncertainty. The informational aspects of Type III uncertainty have been studied for more than six decades and some measures of uncertainty (Klir and Folger, 1988) have been formulated which may be useful to geologists.

Geologists normally are concerned with syntactic, rather than semantic, ignorance and are well aware of their lack of understanding of geological processes. Attempts to handle Type III uncertainty have centered on fuzzy sets (Kacewicz, 1984, 1985, 1987, 1989; Kacewicz and Zydorowicz, 1988; Fang and Chen, 1990). However, objective bases also exist for estimating and reducing Type III uncertainty; for example, various models may be compared (Knopman and Voss, 1988) to determine the superior model and to provide an estimate of the degree of uncertainty. Because so little research has been done on syntactic Type III uncertainty and so little consideration has been given to ignorance, the necessity for assessing this type of uncertainty in nuclear waste repositories provides geologists an excellent opportunity to contribute significantly to its understanding.

Uncertainty from all sources for all three types must be integrated into each final value or prediction. Incorporation of Type I and II uncertainty is relatively easy (ANSI/ASME, 1986; Ronen, 1988; Coleman and Steele, 1989). Methods for including Type III uncertainty, however, are just beginning to be understood, developed, and established.

Current Geologic Use of Uncertainty

Geologic Time Scales. In the seven decades since publication of Barrell's (1917) geologic time scale, at least 25 new scales have been published, of which only seven indicated confidence intervals for numerical values. During the past decade only four of 12 new scales specified uncertainty. Comparison of ages in new time scales with uncertainty values in the immediately preceding geologic time scale shows that 40 percent of all new dates fall outside the estimated limits of uncertainty of its predecessor. Geologic time scales, with a single exception (Barrell, 1917), have been overly optimistic about uncertainty in radiometric dates.

Handbooks of Physical Constants. In the 1942 handbook (Birch *et al.*, 1942), out of 163 tables and 19 figures of data, only two tables indicate relatively complete uncertainties; two tables give limited uncertainties, and 18 tables have rare indications of uncertainties. A subsequent revision (Clark, 1966) contains 224 tables and 105 figures of data, but 94.5% of the tables and figures provide no indication of uncertainty. Interestingly, some indications of uncertainty in the 1942 handbook were deleted in the 1966 handbook, implying that geologists lost knowledge between 1942 and 1966! In both instances in which new values appeared for constants stated with uncertainty in the 1942 handbook, new values in the 1966 handbook were beyond the 1942 limits of uncertainty. Geologists can take heart in knowing that other sciences have similar problems. Reportedly, all 15 new observations of the mean distance to the Sun that were published between 1895 and 1961 fell outside the uncertainty limits of previously estimated distance (MacDonald, 1972).

Probability Density Functions.   The most outstanding geological statements of uncertainty are in those areas that use pdf's to describe their data. A pdf [or an empirical density function (edf) based on observations] completely describes the uncertainty due to Type II stochasticity (Fig. 1). Examples include cumulative distribution functions (cdf's) that sedimentologists may use to determine if two populations of sediments with different origins have been mixed and pdf's, edf's, or cdf's that may be valuable in economic assessment of mineral resources or human safety in risk analysis.

Ranges.   Geologists commonly provide ranges for geological variables such as thicknesses of stratigraphic units or the physical properties for a specific lithology. When coupled with information about means, medians, or modes, they provide a simple but incomplete version of uncertainty. When data are shown graphically with standard statistical summaries (mean, standard deviation, *etc.*), they provide the best expression of Type II uncertainty short of a complete pdf or edf.

Hydrogeology.   During the past decade, hydrologists have added uncertainty estimates to their predictions. In part, this reflects their work with hazardous wastes, but it also comes from a realization that flow forecasts cannot be completely accurate and estimates of uncertainty are essential. Hydrologists have developed practical measures of uncertainty in predicted flow fields, dispersion, and quantity of water flow (Massmann *et al.*, 1991). Use of spatial statistics (Clifton and Neuman, 1982; de Marsily, 1986), Bayesian probabilities (Freeze *et al.*, 1989, 1990; Eslinger and Sagar, 1989), and other innovative methods (Merkhofer and Runchal, 1989) have greatly improved the ability of hydrologists to evaluate uncertainties.

Qualitative Geology.   Uncertainty can be applied to nonquantitative aspects of geology, perhaps best illustrated by the practice of mapping Precambrian outcrops with solid lines and colors to indicate the objective basis for geologic mapping (*e.g.*, Fahrig, 1967). Subjective aspects are indicated by dashed and dotted lines and half-tone colors. This clearly indicates where outcrops may be found and the amount of uncertainty in interpretation can be appraised by the reader.

Comments on Current Practices

"Uncertainty" expressed in a scientific paper usually does not represent estimated total uncertainty but, rather, the statistical precision of the measuring technique (Type I). In geology, even this measure is reported as one standard deviation rather than a recommended two standard deviations (ANSI/ASME, 1986). Rarely is accuracy or bias assessed by comparison with independent means of determining the same value. Almost no effort is made to assess Type II or Type III uncertainties.

All uncertainty should be estimated and cited individually. This point has been made strongly for Type I uncertainty (Eisenhart, 1963) and applies equally well for all other types of uncertainty. Data without estimates of their uncertainty are not worth taking (MacDonald, 1972) and the observations are almost useless for many purposes.

Uncertainty and Mathematical Geologists

Age Dating. McGee and Johnson (1979) developed the first method for standardizing the treatment of experimental errors in fission track dating of rocks. A bivariate normal distribution represents the number of spontaneous fission tracks (Ns), the number of tracks induced by radiation (Ni), and their correlation which diminishes the standard error of Ns/Ni and hence, age error. Neutron flux error is included in the model to further refine the age estimates.

Curl (1988) proposed a Bayesian estimation of ages of rocks by assuming they follow a log normal distribution. Switzer *et al.* (1988) developed a similar model for estimating uncertainty in soil ages by considering the true, but unknown ages to be random variables. Vistelius and Faas (1991) have proposed a more rigorous method for isochron dating that is unbiased, efficient, and consistent. The robust procedure does not assume a Gaussian error distribution and provides confidence intervals which are not necessarily symmetrical.

Geostatistics. One of the outstanding contributions of spatial statistics or geostatistics is that it provides a measure of uncertainty in the regionalized variable (Journel, 1988; Isaaks and Srivastava, 1989; Lajaunie, 1990). By using fuzzy measures of kriging and variogram estimation (Bardossy *et al.*, 1990), the probabilistic and fuzzy uncertainties can be assessed, separating Type I uncertainty from Type II (lack of homogeneity in the geologic materials) and Type III (sample spacing).

Sampling and Mapping. Veneziano and Kitanidis (1982) use a sequential sampling method for contour mapping based on an uncertain function that is general and can be used in other applications. Founded on pseudo-Bayesian second-moment analysis, the method is advantageous when the function has an unknown trend and the random process quantifying prior uncertainty is strongly correlated. Shcheglov (1991) has proposed assessing uncertainty in boundaries of ore bodies mapped from borehole data. Such analyses should accompany all geologic maps, cross-sections, and diagrams. Kacewicz (1991)

Table 2: Modal Analyses (from Stockelmann and Reimold, 1989, Table 3)

| Parameter | Components | | | | |
	Gneiss	Leptynite	Granite	Amphibolite	Diorite
SiO_2	66.39 ± 2.85	75.01 ± 3.15	73.02 ± 2.99	53.60 ± 2.48	59.52 ± 3.00
TiO_2	0.86 ± 0.16	0.20 ± 0.09	0.22 ± 0.22	0.95 ± 0.61	0.83 ± 0.60
Al_2O_3	16.30 ± 1.65	13.67 ± 1.50	14.91 ± 0.98	17.41 ± 1.88	18.62 ± 2.00
Fe_2O_3	5.74 ± 2.41	2.09 ± 1.23	1.71 ± 1.71	9.75 ± 1.42	6.88 ± 2.00
MgO	2.78 ± 0.45	0.61 ± 0.30	0.74 ± 0.81	6.44 ± 2.48	4.09 ± 2.50
K_2O	3.35 ± 1.59	3.48 ± 1.36	5.67 ± 1.38	0.76 ± 0.37	$5.24 \pm N D$
CaO	1.87 ± 0.87	1.26 ± 0.95	0.79 ± 0.51	8.40 ± 0.85	$1.79 \pm N D$
Na_2O	2.92 ± 1.07	3.61 ± 0.65	3.20 ± 0.92	3.53 ± 0.02	$3.02 \pm N D$

N D = not determined

addresses the same problem through tolerance analysis and fuzzy-set theory to arrive at a measure of boundary or shape uncertainty.

Mining. In addition to widely employed geostatistical methods for assessing uncertainty in mining, Kogan (1989) assigns confidence intervals conditional upon the precision of estimates of prospects and their properties. This technique may be applied at all stages of geological exploration and could be extended to mine operation.

Nuclear Waste Disposal. Uncertainties in assessments of high-level radioactive waste repositories have been discussed by many workers (*e.g.*, Bonano and Cranwell, 1988; Craig 1988; Mann, 1988; Hunter and Mann, 1989, 1992) which are important in predicting safety in disposal of long-life nuclear wastes.

Paleontology. Strauss and Sadler (1989) used Bayesian inference to extend classical confidence intervals to estimate local ranges of taxa. This provides a way to assess uncertainty assuming that current knowledge about a taxon is relatively complete (Type III uncertainty is relatively small).

Parameter Estimation. Estimating parameters from erroneous measurements and estimating coefficients of processes when the parameters are inaccessible by direct measurements was addressed by Christakos (1985). Treating parameters as random functions which can be recursively modeled in space or time has several advantages over traditional methods. The estimates may be more accurate, better represent spatial variation, and be obtained with less computational effort.

Petrology. Stockelmann and Reimold (1989) have proposed an *a posteriori* solution of multicomponent mixing that starts from a known mixture of components and predicts the probable initial melt composition. Optimized harmonic least-squares regression estimates the Type I uncertainty associated with each petrologic component (Table 2).

Undiscovered Pool Sizes. Uncertainties in estimates of undiscovered hydrocarbon reservoirs have been estimated by relatively crude methods that rely on assumed frequency distributions (Lee and Wang, 1985; Rice, 1986). Bayesian methods providing a more rigorous, albeit still subjective, basis which becomes less subjective with each iteration has been utilized by Stone (1990) to estimate both the number of undiscovered pools and their uncertainty from the record of discoveries in a basin or play.

Conclusions

Geologists have not been particularly good at recognizing and expressing the uncertainties in their work in the past, although some noteworthy contributions have been made. Probability density functions are used to describe natural variation found in some geological variables and often reveal information that is helpful for understanding the origin of geologic material or predicting new economic deposits. The development of spatial statistics or geostatistics is another example of geology's contribution to the explicit quantification of uncertainty.

Three types of uncertainty are recognized in geologic work: Type I uncertainties arise from measurement processes and consist of both imprecision and inaccuracies (bias); Type II uncertainties are stochasticity inherent in natural phenomena; Type III uncertainties are the result of limited or faulty scientific knowledge, including the inability to model nature accurately.

Both Type I and Type II uncertainty are relatively well understood and best described by probability theory and probability density functions, although measurement bias must be independently assessed. Type III uncertainty is theoretically unknowable, but must be estimated in some manner, perhaps by subjective judgment and use of fuzzy-set theory, possibility theory, and expert opinion.

Geologists are becoming increasingly concerned about uncertainty and now routinely incorporate uncertainty estimates in their calculations and predictions and actively seek better ways for measuring geologic uncertainty.

Acknowledgment

The author greatly appreciates the comments of Joel Massmann and Marek Kacewicz, whose assistance is gratefully acknowledged.

References

ANSI/ASME, 1986, *Measurement Uncertainty, Part 1. Instruments and Apparatus:* American National Standards Institute/American Society of Mechanical Engineers, ANSI/ASME PTC 19.1–1985, New York, 68 pp.

Apostolakis, George, 1990, The concept of probability in safety assessments of technological systems: *Science*, v. 250, p. 1359–1364.

Bardossy, A., Bogardi, I. and Kelly, W.E., 1990, Kriging with imprecise (fuzzy) variograms. I: Theory: *Math. Geol.*, v. 22, p. 63–79.

Barford, N.C., 1985, *Experimental Measurements: Precision, Error, and Truth, 2nd Ed.:* John Wiley & Sons, Chicester, 156 pp.

Barrell, Joseph, 1917, Rhythms and the measurement of geologic time: *Geol. Soc. Amer. Bull.*, v. 28, p. 745–904.

Beers, Yardley, 1957, *Introduction to the Theory of Error, 2nd Ed.:* Addison–Wesley Publ. Co., Reading, Mass., 66 pp.

Birch, Francis, Schairer, J.F. and Spicer, H.C., (eds.), 1942, *Handbook of Physical Constants:* Geol. Soc. Amer., Special Paper 36, 325 pp.

Bonano, E.J. and Cranwell, R.M., 1988, Treatment of uncertainties in the performance assessments of geologic high-level radioactive waste repositories: *Math. Geol.*, v. 20, p. 543–565.

Christakos, George, 1985, Recursive parameter estimation with applications in earth sciences: *Math. Geol.*, v. 17, p. 489–515.

Clark, S.P., Jr., (ed.), 1966, *Handbook of Physical Constants: Rev. Ed.:* Geol. Soc. Amer., Memoir 97, 583 pp.

Clifton, P.M. and Neuman, S.P., 1982, Effects of kriging and inverse modeling on conditional simulation of the Avra Valley aquifer in southern Arizona: *Water Resources Research*, v. 18, p. 1215–1234.

Coleman, H.W. and Steele, W.G., Jr., 1989, *Experimentation and Uncertainty Analysis for Engineers:* John Wiley & Sons, New York, 199 pp.

Cox, L.A., Jr., 1982, Artifactual uncertainty in risk analysis: *Risk Analysis*, v. 2, p. 121–135.

Craig, R.G., 1988, Evaluating the risk of climate change to nuclear waste disposal: *Math. Geol.*, v. 20 , p. 567–588.

Curl, R.L., 1988, Bayesian estimation of isotopic age differences: *Math. Geol.*, v. 20, p. 693–698.

de Marsily, Ghislain, 1986, *Quantitative Hydrology:* Academic Press, New York, 434 pp.

Eisenhart, Churchill, 1963, Realistic evaluation of the precision and accuracy of instrumental calibration systems: *Jour. Research—C. Engineering and Instrumentation*, National Bureau of Standards, v. 67C, no. 2, p. 161–187.

Eisenhart, Churchill, 1968, Expression of the uncertainties of final results: *Science*, v. 160, p. 1201–1204.

Eslinger, P.W. and Sagar, Budhi, 1989, Use of Bayesian analysis for incorporating subjective information, *in* Buxton, B.E., (ed.), *Geostatistical, Sensitivity, and Uncertainty Methods for Ground-Water Flow and Radionuclide Transport Modeling:* Battelle Press, Columbus, Ohio, p. 613–627.

Fahrig, W.S., 1967, *Shabogamo Lake Map Area, Newfoundland, Labrador and Quebec:* Geol. Survey of Canada, Memoir 354, 23 pp.

Fang, J.H. and Chen, H.C., 1990, Uncertainties are better handled by fuzzy arithmetic: *Bulletin Am. Assoc. Petroleum Geologists*, v. 74, p. 1228–1233.

Freeze, R.A, de Marsily, Ghislain, Smith, Leslie and Massmann, Joel, 1989, Some uncertainties about uncertainty, *in* Buxton, B.E., (ed.), *Geostatistical, Sensitivity, and Uncertainty Methods for Ground-Water Flow and Radionuclide Transport Modeling:* Battelle Press, Columbus, Ohio, p. 231–260.

Freeze, R.A., Massmann, Joel, Smith, Leslie, Sperling, Tony and James, Bruce, 1990, Hydrogeological decision analysis: 1. A framework: *Ground Water*, v. 28, p. 738–766.

Gutjahr, A.L., 1992, Hydrology: Chapter 4, *in* Hunter, R.L. and Mann, C. John, (eds.), *Determining Probabilities of Geologic Events and Processes:* Studies in Mathematical Geology No. 4, Oxford Univ. Press, New York, p. 61–80.

Heisenberg, Werner, 1930, The physical principles of the quantum theory, *in* Newman, J.R., (ed.), *World of Mathematics:* Simon & Schuster, New York, p. 1051–1055.

Hunter, R.L. and Mann, C. John, (eds.), 1989, *Techniques for Determining Probabilities of Events and Processes Affecting the Performance of Geologic Repositories, Vol. 1:* Sandia National Laboratories, SAND86–0196 and NUREG/CR–3964, 170 pp.

Hunter, R.L. and Mann, C. John, (eds.), 1992, *Determining Probabilities of Geologic Events and Processes:* Studies in Mathematical Geology No. 4, Oxford Univ. Press, New York, 306 pp.

Isaaks, E.H. and Srivastava, R.M., 1989, *An Introduction to Applied Geostatistics:* Oxford Univ. Press, New York, 552 pp.

Jaech, J.L., 1985, *Statistical Analysis of Measurement Errors:* John Wiley & Sons, New York, 290 pp.

Journel, A.G., 1988, New distance measures: The route toward truly non-Gaussian geostatistics: *Math. Geol.,* v. 20, p. 459–475.

Kacewicz, Marek, 1984, Application of fuzzy sets to subdivision of geological units: Oral presentation, 27th International Geol. Congress, Moscow, *in* Merriam, D.F., (ed.), 1988, *Current Trends in Geomathematics:* Plenum Press, New York, p. 43–56.

Kacewicz, Marek, 1985, *Fuzzy Sets in Engineering Geology:* Mining Symposium, Pribram, Czechoslovakia, conference materials [microfiche].

Kacewicz, Marek, 1987, Fuzzy slope stability method: *Math. Geol.,* v. 19, p. 753–763.

Kacewicz, Marek, 1989, On the problem of fuzzy searching for hard workability rocks in open-pit mine exploitation: *Math. Geol.,* v. 21, p. 309–318.

Kacewicz, Marek, 1991, Shape prediction with a fuzzy uncertainty measure: *Math. Geol.,* v. 23, p. 289–295.

Kacewicz, Marek and Zydorowicz, T., 1988, Application of fuzzy-set theory to the recognition of fault zones: *Sci. de la Terre,* no. 27, p. 217–227.

Klir, G.J. and Folger, T.A., 1988, *Fuzzy Sets, Uncertainty, and Information:* Prentice–Hall, Englewood Cliffs, N.J., 341 pp.

Knopman, D.S. and Voss, C.I., 1988, Discrimination among one-dimensional models of solute transport in porous media: Implications for sampling design: *Water Resources Research,* v. 24, p. 1859–1876.

Kogan, R.I., 1989, Interval estimation of mineral prospects: *Math. Geol.,* v. 21, p. 607–618.

Lajaunie, C., 1990, Comparing some approximate methods for building local confidence intervals for predicting regional variables: *Math. Geol.,* v. 22, p. 123–144.

Laslett, G.M. and Sandland, R.L., 1989, Precision and accuracy of kriging estimators with interlaboratory trial information, *in* Armstrong, M., (ed.), *Geostatistics, Vol. 2:* Kluwer Academic Publishers, Dordrecht, p. 797–808.

Lee, P.J. and Wang, P.C.C., 1985, Prediction of oil and gas pool sizes when discovery record is available: *Math. Geol.,* v. 17, p. 95–113.

MacDonald, J.R., 1972, Are the data worth owning?: *Science,* v. 176, p. 1377.

Mann, C. John, 1988, Methods for probabilistic assessments of geologic hazards: *Math. Geol.,* v. 20, p. 589–601.

Massmann, Joel, Freeze, R.A., Smith, Leslie, Sperling, Tony and James, Bruce, 1991, Hydrogeological decision analysis: 2. Applications to ground-water contamination: *Ground Water,* v. 29, p. 536–548.

McGee, V.E. and Johnson, N.M., 1979, Statistical treatment of experimental errors in the fission track dating method: *Math. Geol.,* v. 11, p. 255–268.

Merkhofer, M.W. and Runchal, A.K., 1989, Probability encoding: Quantifying judgmental uncertainty over hydrologic parameters for basalt, *in* Buxton, B.E., (ed.), *Geostatistical, Sensitivity, and Uncertainty Methods for Ground-Water Flow and Radionuclide Transport Modeling:* Battelle Press, Columbus, Ohio, p. 629–648.

OECD, 1987, *Uncertainty Analysis for Performance Assessments of Radioactive Waste Disposal Systems:* Organization for Economic Cooperation and Development, Paris, 258 pp.

Rice, D.D., (ed.), 1986, *Oil and Gas Assessment—Methods and Applications:* Am. Assoc. Petroleum Geologists, Studies in Geology No. 21, 263 pp.

Ronen, Yigal, 1988, *Uncertainty Analysis:* CRC Press, Inc., Boca Raton, Florida, 275 pp.

Shcheglov, V.I., 1991, Errors in determining boundaries of stratified deposits: *Math. Geol.,* v. 23, p. 77–85.

Smithson, Michael, 1989, *Ignorance and Uncertainty, Emerging Paradigms:* Springer–Verlag, New York, 367 pp.

Stockelmann, D. and Reimold, W.U., 1989, The HMX mixing calculation program: *Math. Geol.*, v. 21, p. 853–860.

Stone, L.D., 1990, Bayesian estimation of undiscovered pool sizes using the discovery record: *Math. Geol.*, v. 22, p. 309–332.

Strauss, David and Sadler, P.M., 1989, Classical confidence intervals and Bayesian probability estimates for ends of local taxon ranges: *Math. Geol.*, v. 21, p. 411–427.

Switzer, P., Harden, J.W. and Mark, R.K., 1988, A statistical method for estimating rates of soil development and ages of geologic deposits: A design for soil–chronosequence studies: *Math. Geology*, v. 20, p. 49–61.

US EPA, 1985, *Environmental Standards for the Management and Disposal of Spent Nuclear Fuel, High-Level and Transuranic Radioactive Wastes, Final Rule:* U.S. Environmental Protection Agency, 40 CFR Part 191: Federal Register, v. 50, no. 182, p. 38066–38089.

Veneziano, Daniele and Kitanidis, P.K., 1982, Sequential sampling to contour an uncertain function: *Math. Geol.*, v. 14, p. 387–404.

Vistelius, A.B. and Faas, A.V., 1991, On the precision of age measurements by Rb–Sr isochrons in nontrivial cases: *Math. Geol.*, v. 23, p. 999–1044.

21

EXPERT SYSTEMS IN
ENVIRONMENTAL GEOLOGY

W. Skala and S. Heynisch

Geoscientific environmental research deals with very complex problems requiring integration of knowledge from many different fields. Computer-based risk evaluation methods are required as practical tools to automatically compare and evaluate contaminated sites. ALTRISK and HYDRISK are two examples of knowledge-based systems useful for environmental analysis.

Environmental Research and Risk Assessment

What types of problems must be solved by the geosciences that are relevant to environmental issues? Is it possible to define a special field of "environmental geology" that is distinct from the other branches of the geosciences? Geoscientists have always believed that it is part of their responsibility to find solutions to environmental problems. These questions have been treated for many years in numerous publications and discussions, and have been considered in educational programs.

The environmental sciences are characterized by pragmatism. Research must be restricted to that which is sensible and leads to further decisions and actions that result in solutions which are acceptable with respect to their consequences. There are two essential aspects of environmental research: 1) Investigators should use scientific, empirically reproducible methods; and 2) Research should be driven by the significance and consequences of any potential risk—for example, the environmental impact on human health. These aspects cannot be considered separately.

Pragmatic considerations determine the modality and resolution of the research. This has the effect of "styling" the mode of operation and the approach to specific problems. The method of sampling influences the perception of effects in a specific situation and also determines the assessment

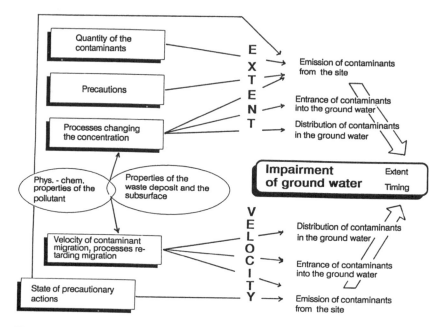

Figure 1: Factors involved in impairment of ground water by emissions from a contaminated site.

of risk. Because of the very complex correlations between many causes, risk cannot be determined in an exact manner. It is necessary that an expert be asked for an opinion of the degree of risk involved, based on his or her past experience in similar circumstances.

Environmental research requires interdisciplinary investigations and the integration of knowledge from different fields. Even taking into account all of the known interrelationships, an exact evaluation of the risk may be impossible. The very complex problems dealt with by geoscientific environmental research are characterized by many interactions that depend upon conditions within the system. Such complexity is illustrated for the problem of ground-water contamination in Figure 1.

The comparison of different potential risks and the ranking of potentially hazardous sites is more useful for drawing conclusions than attempting to assess an individual site independently. However, comparability is a necessity and requires that equivalent methods and equivalent rules or standards be used in the evaluation of each situation. Investigations and evaluations must be based on explicit procedures and observations and the investigator should be able to duplicate individual steps in an assessment.

It also should be possible to evaluate potential risk in a dynamic manner in which changing conditions of space and time are taken into account. This is especially important if forecasts must be made for a developing environmental situation, such as the progressive contamination of a productive aquifer that supplies potable water for a community.

Expert Systems in Environmental Geology

Expert systems provide a structure for support of environmental assessment that fulfills the requirements stated above. Knowledge-based systems do not replace the expert, but rather support the expert's work. Investigation and evaluation of potential environmental risks is a continuing process; therefore, it should be possible to draw successive conclusions from information in the database. Because of pragmatic limitations on time and investigative resources, it is impossible for the scientist to analyze a troublesome situation perfectly. However, one must only solve the question as far as necessary to support the making of a particular decision. In such cases, statistical methods, and in the case of spatial properties, geostatistical methods, must be used.

These fundamental considerations will be illustrated with several studies currently being carried out at the Institute for Geoinformatics at the Free University of Berlin. The primary aim of these studies is to create knowledge-based systems to evaluate the potential risk to ground water from contamination by waste dumps or industrial sites.

ALTRISK—A Knowledge-Based System for Risk Recognition

The expert system ALTRISK (Osterkamp, Richter and Skala, 1989a) evaluates potential risk to ground water, based on an initial investigation of the site. The program's goal is to support investigators in judging contaminated sites. ALTRISK provides quantified risk evaluation and offers the possibility of comparing different sites. It provides tools for setting priorities according to either remediation methods or site characteristics.

ALTRISK asks for information and knowledge about the specific event being investigated. The knowledge is obtained using an interactive dialogue that provides four choices: 1) Evaluation of waste disposal sites, 2) Evaluation of separate criteria (scenarios), 3) Evaluation of separate criteria to select a new location, and 4) Aids to assess priorities.

In a subsequent step, information provided by the user is evaluated and transformed into risk values. Finally, the system issues a report containing the individual evaluation of each step that has influenced the final assessment of risk. Therefore, it is possible to follow the successive degree of risk along the pathway: *contaminated sites → ground water → human consumption* and to propose suitable precautionary measures. Follow-up steps might include additional investigations or preventive actions at the contaminated site. If critical information is missing or the database is insufficient, the system requests further investigations. If the data are sufficient, the realiability of the calculated risk reflects the significance of the input data.

ALTRISK has one important advantage over conventional methods of risk assessment; it is capable of analyzing several levels of investigation. In addition to identification and evaluation of risk, scenarios can be run to check the hypothetical success of different methods of remediation. The reliability of an evaluation depends primarily on the nature and origin of the input data. At the beginning of an evaluation of a contaminated site, there generally is little information. At this stage statements about the potential risk can be made, but they are not very precise. As the level of investigation increases, the reliability of the evaluation (risk assessment) improves, but a minimum amount of data and information is necessary.

The program recognizes four levels of information—A: Memoranda, records, testimonies, and publications; B: Records, other documents, and field checkings; C: Records, other documents, and first field investigations; D: First field investigations and drill logs. The criteria for risk assessment must be modified by weighting factors according to relevance and influence. If the information available is incomplete, the weighting of criteria leads to different consequences. At the extreme, the absence of information concerning an important criterion may stop the evaluation. In other circumstances, it may be possible to use empirical default values, rather than observed values.

General climatic data are usually available, such as precipitation records measured over long periods in a certain region. Such averaged measurements can be used in the evaluation system. The reliability of regional data is, of course, not as high as rainfall measurements made in the immediate neighborhood of the site under investigation. Experience has shown that the use of regional data may lead to problems, the severity of which should not be underestimated.

Investigations of waste deposits have demonstrated that a risk assessment for ground water leads to false estimations if it is based only on chemical analyses. Although chemical, physical, and biological factors play important roles in the transport of contaminants, the effects of dilution, velocities of ground-water movement, and prior abnormal concentrations of contaminants should not be overlooked. A final evaluation of ground-water quality cannot be made before a sufficient number of samples are examined. Additional investigations are necessary for reliable risk assessment of ground-water contamination.

Risk Assessment of Subsurface Conditions at Contaminated Sites

Until now, sufficient geological and hydrological criteria for a thorough assessment of contaminated sites have not been available. The complicated

1	2	3	4
ALTRISK	HYDRISK		
A B C D			
A: Memoranda, records, testimonies, and publications.	Aim: Recognition and identification of conspicuous changes in the environment.	Aim: Overview of type and extent of source of danger.	Aim: Precise demarcation of source of danger to the environs; decision on suitable remedial actions.
B: Records, other documents, and field checkings.	Recognition of pollutants.		
C: Records, other documents, and first field investigations.	Chemical analyses.	Chemical analyses related to a narrow screen.	Chemical analyses related to a screen.
D: First field investigations and drill logs.	Simple subsurface investigation.	Subsurface investigation related to a screen.	Technical methods related to a narrow screen; contaminated transport modeling; tracer and pump methods.

Figure 2: Levels of site information available and corresponding appropriate levels of investigation incorporated in ALTRISK and HYDRISK.

regional distributions of geological media and chemical processes within the unsaturated zone and in ground water have not been well defined, causing problems in the application of expert systems. In many countries such as Germany, the pertinent investigations, as well as the planning and management of remedial actions, proceeds in a step-by-step manner (Fig. 2). Generally, no spatially oriented investigation of chemical, geological, or hydrological measurements forms a part of the first step in environmental assessment. The initial information level required by the knowledge-based system ALTRISK consists mostly of global observations of the contaminated site and its environment, plus some initial indications of the pollutants and subsurface conditions.

HYDRISK—A Knowledge-Based System for Spatial Recognition of Risk

A project has been undertaken to develop a risk-assessment system based on a high level of investigation; *i.e.*, on a spatial net of boreholes yielding extensive sample data. The project's aim is development of a knowledge-based system for investigating and evaluating hydrogeological criteria relevant to contaminant transport for the purpose of recommending remedial action. A risk evaluation based on such a high level of information enables the scientist to offer differentiated recommendations about remediation methods useful at the contaminated site. Possible future land uses in the contaminated area are also taken into account.

In contrast to ALTRISK, the basic approach used in HYDRISK emphasizes the spatially dependent and time-dependent aspects of contaminant transport (Heynisch et al., 1992). Most existing expert systems, including risk-assessment systems, are focused mainly on the lowest level of investigation, and neglect the hydrological processes in the geological media and the interactions of contaminants. Surface water and ground water are the main transport media for pollutants. Therefore, it is essential to know the geological and hydrological properties in an area if risk assessment is to be successful (Fritz and Dörhöfer, 1988). Complexities of the three-dimensional variability of geological structures and hydrological processes in space and time forces us, as a first step, to evaluate only the most distinct properties of the subsurface which are critical in contaminant transport in ground water. These factors include the thickness of the unsaturated zone, the transport rate, and the transport direction of a contaminant. Only a thorough hydrological evaluation provides indications that potential hazards to humans and/or the ecological system exist.

Parameters necessary to estimate the danger of pollutants infiltrating the ground water, such as the heterogeneity of the geology and the hydraulic properties of the unsaturated zone, may not be taken into account adequately in an expert system. Low-permeability rocks such as clays, for example, are often considered as reliable barriers to subsurface fluid movement. However, clay layers may exhibit enormous changes in hydrological conductivity because of connected fractures, to the extent that the infiltration of contaminants increases immensely. The regional distributions of highly complicated aquifer systems with different hydraulic conductivities must be described in order to assess ground-water contamination. The most important parameters are hydraulic conductivity, hydraulic head distribution, and hydraulic gradient. Only with sufficient information about these parameters can a potential hazard be described satisfactorily.

In general, the heterogeneity of an aquifer cannot be described in detail because of the cost and time required for the necessary investigations. Unfortunately, transport calculations based on assumptions of homogeneous subsurface conditions lead to erroneous estimates of risk. One possible approach to the definition of heterogenity in the subsurface is the factor of dispersivity. If the dispersion properties of an aquifer can be estimated, a possible distribution of waste concentrations in the aquifer can also be estimated (Schafmeister–Spierling and Pekdeger, 1989).

After the magnitude of initial pollution at the source, the mobility and persistence of a contaminant is the second most important factor. Therefore, the estimated spatial distribution and the order of magnitude of contamination in an area is complicated by factors that control the mobility and persistence of the pollutant. Although the general characteristics of the major pollutants of ground water are reasonably well known, their persistences and mobilities may deviate from some published standards because of physical, chemical, and microbiological conditions in the subsurface. In the case of nonorganic components, the chemical composition of ground water, which is primarily determined by the geological environment, is also important. Metallogenic anomalies, for instance, may lead to high values of heavy metals available to the ground water; on the other hand, the solubility of heavy metals is controlled by the physico-chemical environment. It can be inferred that local deviations from the expected (mean) values of fluid-flow properties and of concentrations of potential contaminants may result in erroneous risk estimates for waste disposal sites. Such a possibility should be excluded for the obvious reason that remedial actions are expensive.

A spatially dependent/time-dependent investigation of geological and hydrological properties in a given area can be performed economically and efficiently by computer. Therefore, it is obviously advantageous to perform the risk assessment of contaminated sites using a knowledge-based system. An appropriate database and information system should enable the user to: 1) recall stored data, 2) represent the spatial distributions of the data, 3) consult stored expert knowledge, and 4) calculate a risk assessment based on a spatial consideration of the geological and hydrological situation.

The prototype of the risk-evaluation system HYDRISK was presented in November 1991 in Berlin at the first meeting of the German Working Group for Mathematical Geology and Geoinformatics. The focal point of this system is the ground-water path which must be protected. The subsurface is divided into different levels: soil, geological layers in the unsaturated zone, and the aquifer itself—which also may consist of layers characterized by different properties. A pollutant may be considered as entering from different starting positions (Fig. 3). First, the pollutant could already have reached the aquifer and contaminated the ground water to varying degrees. In this situation it is very important to determine the potential risk of the use of the contaminated ground water. Second, there may be contaminants in the soil, but the pollutants may not yet have reached the water table. There is a high probability that the ground water will be contaminated in the future. Here, the question of potential risk to the aquifer has to be answered.

Both the diagnostic path and the prognostic path are taken into account in the knowledge-based HYDRISK system. Additionally, HYDRISK has the advantage of a modular structure, to the extent that such can be constructed in an expert system. It therefore represents a tool for evaluating a specific subsurface site by its properties in order to predict the risk to ground water in the case of a hypothetical accident (the so-called German "UVP-Investigation"), and in addition it supports experts seeking to determine

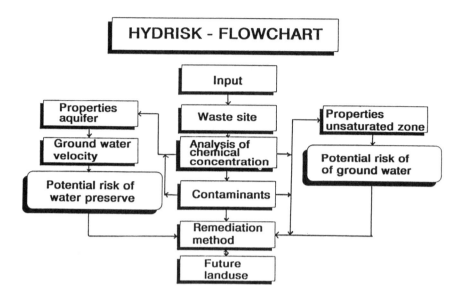

Figure 3: Flow chart for HYDRISK.

the environmental significance of a contaminated site. The system evaluates the properties of the subsurface and their relation to the chemical and physical properties of the pollutants as these affect migration and transport processes (Darimont, 1985). For remediation conceptions, support of the system by statistical and geostatistical algorithms as well as numerical models to calculate the fluid flow and transport properties and to estimate possible contributions of contaminants to the ground water is useful.

Risk Evaluation

In the present context, "risk" is the probability that ground water will be polluted so that land use or ground-water use is or will be dangerous. Because the sites and the geologic and hydrologic properties are geographically distributed, the evaluation of risk is spatially related. Analyses of the knowledge base lead to two types of objects, the so-called "semantic objects" and "syntactic objects." Semantic objects are described by factors which influence the risk of contamination of the ground water and are elucidated by questions about which factors may have caused soil and ground-water contamination. Syntactic objects are space related and represent a spatial unit which is being evaluated. Each syntactic object is defined by its homogeneous semantic attributes. During evaluation, each spatial unit (polygon,

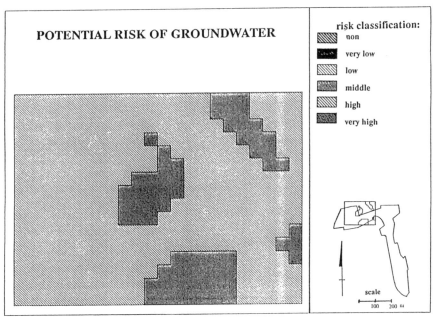

Figure 4: Computer screen visualization of the spatial distribution of potential risk to ground water.

grid cell, *etc.*) is assessed for its associated attributes and is ranked. The various attributes are then replaced by risk values.

Combination with a GIS System

The knowledge-based system must be supported by a geographic information system (Schilcher and Fritsch, 1989) which enables the spatially distributed properties and results of evaluations to be visualized (Fig. 4). An interactive component of the system gives the user the ability to see and check results of the spatial evaluation and helps to make the expert system transparent to the user. In addition, the GIS system offers various spatial operations which help the user obtain space-related information for the risk evaluation. One task, for instance, consists of dividing an area of investigation into homogeneous units such as grid cells or polygons which are to be evaluated. A GIS system also has the capability of combining or intersecting different spatial objects. Homogenous intersections which have the same contents can be combined into a single polygon.

Interaction between the expert system and the GIS system is conducted by file transfer. The knowledge-based system orders a certain operation by sending a specific batch file to the GIS system, which executes the operation and sends the results back to the expert system. Here further evaluation continues, based on the spatial data that has been received.

Conclusions and Outlook

Important experience has already been gained using the programs ALTRISK and HYDRISK during development of the systems. The very detailed preparation of the knowledge base and production of the computer software have systematically elucidated new fields for future research. HYDRISK will be tested on industrial sites using available data. Both systems are being used in several studies for determining the urgency of remediation and choosing suitable remedial actions. Optimization of analysis and sampling methods is being investigated, as well as optimization of the density of spacially dependent and time-dependent sampling (Teutsch *et al.*, 1990).

References

Darimont, T., 1985, Naturwissenschaftlich-technische Anforderungen an die Sanierung kontaminierter Standorte, Teil I: Erarbeitung eines Schemas zur Bewertung des Migrationsverhaltens von Stoffen im Untergrund: *Umweltforschungsplan des Bundesministers des Inneren, Wasserwirtschaft UBA Forschungsbericht 102 03 405/01*, Fachbereich Synthetische und Analytische Chemie der TU Berlin, Berlin, 149 pp.

Fritz, J. and Dörhöfer, G., 1988, Geowissenschaftliche Rahmenuntersuchungen zur Bewertung von Altablagerungen auf Landesebene: *Altlastensanierung '88, 2. Internationaler TNO/BMFT-Kongreß über Altlastensanierung*, Bd. 1, Bundesminister für Forschung und Technologie, Bonn, und Umweltbundesamt, Berlin, p. 217–219.

Heynisch, S., Pekdeger, A., Richter, B. and Skala, W., 1992, HYDRISK—Ein wissensbasiertes System zur Bewertung hydrogeologischer Parameter beim Schadstofftransport im Hinblick auf Sanierungsmaßnahmen: *Beiträge zur Mathematischen Geologie und Geoinformatik*, Bd. 4, Verlag Sven von Loga, Köln, p. 31–39.

Osterkamp, G., Richter, B. and Skala, W., 1989*a*, ALTRISK—Ein wissensbasiertes System zur Bewertung des Gefährdungspfades Grundwasser: *Proc. of* ENVIROTECH, Wien, p. 11.

Osterkamp, G., Richter, B. and Skala, W., 1989*b*, Anforderungen an ein wissensbasiertes System zur Bewertung von Gefährdungspotentialen, *in* Jaeschke, A., Geiger, W. and Page, B., (eds.), *Informatik im Umweltschutz*: Proc. 4. Symp., Nov. 6–8, Karlsruhe, p. 395–405.

Schafmeister–Spierling, M.-T. and Burger, H., 1989, Spatial simulation of hydraulic parameters for fluid flow and transport models, *in* Armstrong, M., (ed.), *Geostatistics*, Vol. 2: Kluwer Academic Publ., Dordrecht, p. 629–638.

Schafmeister–Spierling, M.-T. and Pekdeger, A., 1989, Influence of spatial variability of aquifer properties on groundwater flow and dispersion, *in* Kobus, H.E. and Kinzelbach, W., (eds.), *Contaminant Transport in Groundwater*: Balkema, Rotterdam, p. 215–220.

Schilcher, M. and Fritsch, D., (eds.), 1989, *Geo-Informationssysteme: Anwendung neue Entwicklungen*: Wichmann, Karlsruhe, 364 pp.

Teutsch, G., Ptak, T.H., Schad, H. and Deckel, H.-M., 1990, Vergleich und Bewertung direkter und indirekter Methoden zur hydrologischen Erkundung kleinräumig heterogener Strukturen: *Z. dt. geol. Ges.*, Bd. 141, Hannover, p. 376–384.

22

FROM MULTIVARIATE SAMPLING TO THEMATIC MAPS WITH AN APPLICATION TO MARINE GEOCHEMISTRY

J.E. Harff, R.A. Olea and G.C. Bohling

Integration of mapped data is one of the main problems in geological information processing. Structural, compositional, and genetic features of the Earth's crust may be apparent only if variables that were mapped separately are studied simultaneously. Geologists traditionally solve this problem by the "light table method." Mathematical geologists, in particular, D.F. Merriam, have applied multivariate techniques to data integration (Merriam and Sneath, 1966; Harbaugh and Merriam, 1968; Merriam and Jewett, 1988; Merriam and Sondergard, 1988; Herzfeld and Merriam, 1990; Brower and Merriam, 1990). In this article a regionalization concept based on the interpolation of Bayes' probabilities of class memberships is described using a geostatistical model called "classification probability kriging."

The Regionalized Classification Method

The problem of interpolation between data points has not been considered in most of the publications on multivariate techniques mentioned above. An attempt at data integration—including interpolation of multivariate data vectors—was made by Harff and Davis (1990) using the concept of regionalized classification. This concept combines the theory of classification of geological objects (Rodionov, 1981) with the theory of regionalized variables (Matheron, 1970; Journel and Huijbregts, 1978). The method is based on the transformation of the original multivariate space of observed variables into a univariate space of rock types or rock classes. Distances between multivariate class centers and measurement vectors within the feature space are needed for this transformation. Such distances can be interpolated between the data points using kriging. Because of the assumptions of multinormality and the fact that Mahalanobis' distances tend to follow a χ^2 distribution, the distances must be normalized before kriging (Harff, Davis and Olea, 1991). From the resulting normalized distance vectors at each node of a spatial grid, the Bayes' probability of class membership can be calculated for each class. The corresponding grid nodes will be assigned to the classes with the greatest membership probabilities. The result is a regionalization scheme covering the area under investigation.

Let $X(r)$ denote the multivariate field of features, modeled as a regionalized variable (RV). $I = \{1, \ldots, K\}$ denotes a discrete variable indicating K different rock types, and R denotes the region of investigation, with each point described by an x–y-coordinate vector, r. The classical geological mapping problem can be described as a mapping, M, of the multivariate field of geological features within the region onto the field of rock types or units

$$M : X(r) \longrightarrow I(r), \forall r \epsilon R. \tag{1}$$

For the random field $X(r)$, the realization is given as vectors of values $x(r_j)$ at a limited number of sampling points $r_j \epsilon R$, $j \epsilon J$, $J = \{1, \ldots, N\}$. These measurement vectors are allocated to classes, each of which stands for one of the rock types denoted by I. That means that a mapping M' is given for the measurement points

$$M' : \{x(r_j)\}_{j \epsilon J} \longrightarrow I. \tag{2}$$

Based on this classification, a matrix $\mathbf{B}$ of unstandardized discriminant coefficients and a vector $\mathbf{a}$ of associated constants are calculated (Fisher, 1936; Tatsuoka, 1971). Using $\mathbf{B}$ and $\mathbf{a}$, a transformation can be obtained that will provide vectors f of new random multivariate variables F

$$f(r_j) = X(r_j)\mathbf{B} + \mathbf{a}, \forall j \epsilon J. \tag{3}$$

For each class, i, of the K classes, a mean vector c_i^* and a covariance matrix D_i must be estimated for the new variables, F. Mahalanobis' distance between the transformed measurement vector, $f(r_j)$, and each transformed group centroid, c_i^*, is given by

$$\chi_i^2(r_j) = (f(r_j) - c_i^*)'D_i^{-1}(f(r_j) - c_i^*). \tag{4}$$

Bayes' probability of class membership is determined using the distances $\chi_i^2(r_j)$

$$p_i(r_j) = \frac{p_i' \mid D_i \mid^{-1/2} \exp(-\chi_i^2(r_j)/2)}{\sum_{k \epsilon I} p_k' \mid D_k \mid^{-1/2} \exp(-\chi_k^2(r_j)/2)} \tag{5}$$

where p_i' stands for the *a priori* probability of class i. It is assumed that the spatial correlation and ergodicity of RV $X(r)$ are preserved through the transformations and that the probabilities $p_i(r_j)$ can be regarded as realizations of an RV $P_i(r)$,

$$P_i(r) = m_i(r) + Y_i(r), \forall r \epsilon R, \tag{6}$$

where $m_i(r)$ denotes expected value and $Y_i(r)$ is a stochastic fluctuation.
The semivariogram, $\gamma_i(h)$, measures spatial correlation

$$\gamma_i(h) = 1/2E[(Y_i(r+h) - Y_i(h))^2], \tag{7}$$

where h denotes a distance in R. Using ordinary kriging, probabilities are
estimated at each node $r' \epsilon R'$, on a spatial grid $R' \subset R$

$$p_i^*(r') = \sum_{j \epsilon V} \lambda_j p_i(r_j), \forall r' \epsilon R' \tag{8}$$

using probabilities calculated at measurement points r_j within the vicinity
V, $V \subset J$, of r' (Harff, Davis, Olea and Bohling, 1991). This yields the
probability vectors at each grid node for each class $\forall i \epsilon I$,

$$\begin{pmatrix} p_1^*(r') \\ p_2^*(r') \\ \vdots \\ p_K^*(r') \end{pmatrix} \forall r' \epsilon R'. \tag{9}$$

The regionalization required in Eq. (1) may now be determined:

$$i(r') = \{i \epsilon I : p_i^*(r') = \overset{\max}{k} \ p_k^*(r')\}, \forall r' \epsilon R'. \tag{10}$$

A reliability function for the regionalization can be given as

$$p^*(r') = \{p_i^*(r') \epsilon P^*(r') : p_i^*(r') = \overset{\max}{k} \ p_k^*(r')\}, \forall r' \epsilon R' \tag{11}$$

where $P^*(r')$ is the set of elements in vector Eq. (9). In other words, each
grid node is assigned to the class that has the maximum probability at that
node. The map of maximum probabilities serves as a guide to the reliability
of class assignments. If the maximum value among the class membership
probabilities is low, the classification at that location is uncertain. Typically,
"valleys" of low maximum probability occur along the boundaries between
subregions assigned to different classes.

Case Study

Chemical compositions of manganese nodules collected from the floor of the
Central Pacific Ocean between 160°W to 105°W and 0° to 30°N were used
for a case study. Harff, Lange and Olea (1991) had classified the manganese
nodules on the basis of seven metals: manganese, iron, cobalt, nickel, copper,
zinc and lead. The survey includes 99 observations resulting from averaging

Table 1: Means of Metal Content (%) in Types of Manganese
Nodules from Pacific Ocean (after Harff, Lange and Olea, 1991)

Dendrogram	Type	Mn	Fe	Co	Ni	Cu	Zn	Pb
	Ia	27.85	6.23	0.212	1.410	1.202	0.146	0.044
	Ib	25.39	10.00	0.147	1.093	0.607	0.187	0.038
	II	22.87	7.85	0.269	1.142	0.913	0.093	0.075
	III	18.34	12.29	0.206	0.643	0.426	0.066	0.842

geochemical data from manganese nodules sampled by drilling and dredging
at contiguous locations.

The data vectors were classified by a hierarchical agglomerative cluster analysis. We distinguish four classes, each representing a nodule type.
Table 1 lists the mean metal content of the types; a schematic dendrogram
indicates the approximate affinities between the types. On the basis of their
metal content, the manganese nodule types can be characterized as:

Mn/Ni/Cu–type (Ia)   Mn/Zn–type (Ib)
Co(Ni/Cu)–type (II)   Fe/Pb–type (III).

Bayesian probabilities were calculated for the classification of every grid
node into each nodule type. Semivariograms were estimated assuming quasi-
stationarity (Neuman and Jacobson, 1984), using the coordinates of the
centers of the sampling areas, which can be regarded as points at the scale
of the study. Figures 1 to 4 show the experimental semivariogram values
and the selected semivariogram models.

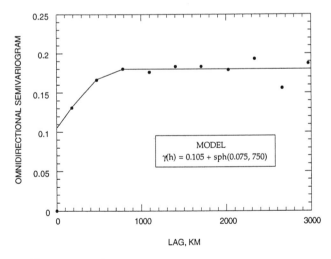

Figure 1: Experimental semivariogram and spherical model for class membership probability [Eq. (5)], Mn/Ni/Cu–type (Ia).

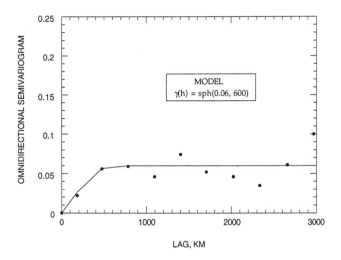

Figure 2: Experimental semivariogram and spherical model for class membership probability [Eq. (5)], Mn/Zn–type (I*b*).

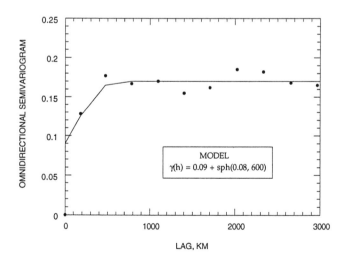

Figure 3: Experimental semivariogram and spherical model for class membership probability [Eq. (5)], Co(Ni/Cu)–type (II).

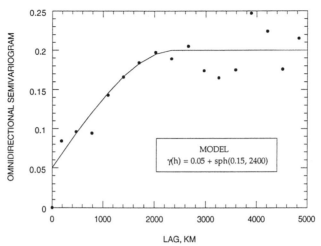

Figure 4: Experimental semivariogram and spherical model for class membership probability [Eq. (5)], Fe/Pb–type (III).

It is clear that correlation ranges of the class membership probabilities are larger than the ranges of individual metal contents (Harff, Lange and Olea, 1991). The increase in the range results from data integration and generalization. Interpolation was done using ordinary kriging (Journel and Rossi, 1989). Figures 5 to 8 show class membership probabilities as isoline maps. The regionalization scheme based on estimated class membership probabilities is given in Figure 9. Figure 10 is a probability map [Eq. (11)] which expresses the reliability of the regionalization.

Summary

Classification probability kriging allows the integration of multivariate data measured at a limited number of sampling points into a thematic map covering the entire region of investigation. The method is implemented in four steps:

1. Vectorial measurements are classified and the classes interpreted as representations of different nodule types,

2. Bayes' probabilities of class membership at each measurement point are calculated,

3. Class membership probabilities are estimated by kriging at each node of a regular grid,

4. Each grid node is assigned to the class that has the greatest membership probability at the node.

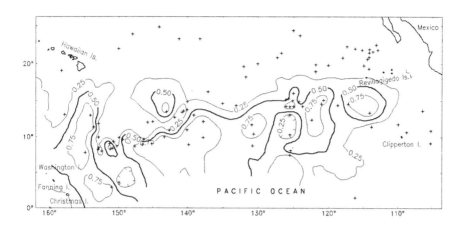

Figure 5: Class membership probability [Eq. (5)] of the manganese nodules Mn/Ni/Cu–type (I*a*) in the central Pacific Ocean.

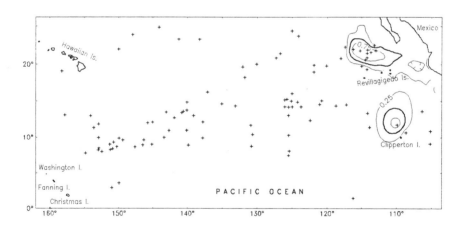

Figure 6: Class membership probability [Eq. (5)] of the manganese nodules Mn/Zn–type (I*b*) in the central Pacific Ocean.

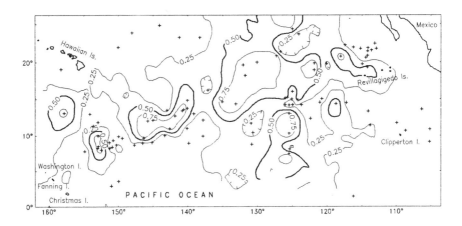

Figure 7: Class membership probability [Eq. (5)] of the manganese nodules Co(Ni/Cu)–type (II) in the central Pacific Ocean.

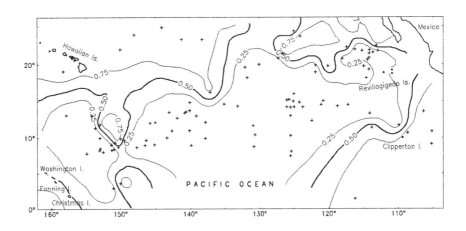

Figure 8: Class membership probability [Eq. (5)] of the manganese nodules Fe/Pb–type (III) in the central Pacific Ocean.

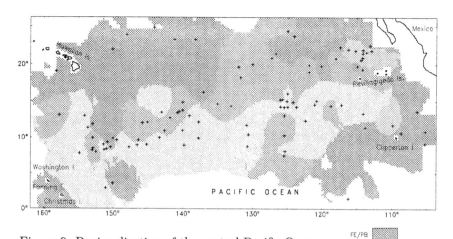

Figure 9: Regionalization of the central Pacific Ocean
sediments expressing the distribution of manganese
nodule types Ia, Ib, II, and III.

FE/PB

CO(NI/CU)

MN/ZN

MN/NI/CU

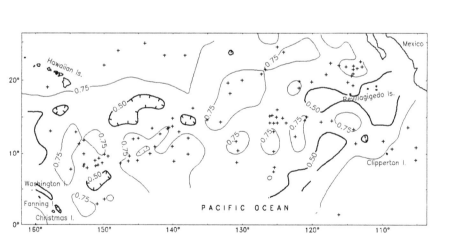

Figure 10: Reliability of the regionalization in Figure 9 expressed as probability according to Eq. (11).

The procedure can be interpreted as a reduction of the multivariate space of measured variables to a univariate space of classes (nodule types). Application of the method to the distribution of manganese nodules in the central Pacific Ocean distinguishes four nodule types on the basis of seven metals: type Ia (Mn/Ni/Cu), type Ib (Mn/Zn), type II (Co(Ni/Cu)), type III (Fe/Pb). The class membership probabilities show general east–west trends.

Acknowledgments

The authors thank Prof. Dr. D. Lange, Institute for Baltic Sea Research, Warnemünde, Germany, for providing the data. We also thank Dr. David R. Collins and Richard L. Brownrigg of the Kansas Geological Survey for cartographic assistance. Research was supported by the German Science Foundation and the Kansas Geological Survey.

References

Brower, J.C. and Merriam, D.F., 1990, Geologic map analysis and comparison by several multivariate algorithms, in Agterberg, F.P. and Bonham–Carter, G.F., (eds.), Statistical Applications in the Earth Sciences: Geol. Survey of Canada, Paper 89–9, p. 123–134.

Fisher, R.A., 1936, The use of multiple measurements in taxonomic problems: Annals of Eugenics, v. 7, no. 2, p. 179–188.

Harbaugh, J.W. and Merriam, D.F., 1968, Computer Applications in Stratigraphic Analysis: John Wiley & Sons, New York, 282 pp.

Harff, J. and Davis, J.C., 1990, Regionalization in geology by multivariate classification: Math. Geol., v. 22, no. 5, p. 573–588.

Harff, J., Davis, J.C. and Olea, R.A., 1991, Quantitative assessment of mineral resources with an application to petroleum geology: Nonrenewable Resources, v. 1, no. 1, p. 74–84.

Harff, J., Davis, J.C., Olea, R.A. and Bohling, G., 1991, Regionalization of Western Kansas Based on Multivariate Classification of Stratigraphic Data from Oil Wells, II: Kansas Geol. Survey, Open-File Report 91–40, 30 pp.

Harff, J., Lange, D. and Olea, R.A., 1991, Geostatistics for computerized geological mapping: Geol. Jb., v. A 122, p. 473–483.

Herzfeld, U.C. and Merriam, D.F., 1990, A map comparison technique utilizing weighted input parameters, in Gaal, G. and Merriam, D.F., (eds.), Computer Applications in Resource Estimation: Pergamon Press, Oxford, p. 43–52.

Journel, A.G. and Huijbregts, C.J., 1978, Mining Geostatistics: Academic Press, New York, 600 pp.

Journel, A.G. and Rossi, M.E., 1989, When do we need a trend model in kriging?: Math. Geol., v. 21, no. 7, p. 715–739.

Matheron, G., 1970, La Théorie des Variables Régionalisées, et ses Applications: Les Cahiers du Centre de Morphologie Mathématique de Fontainebleau, Fascicule 5, Ecole Nationale Supérieure de Mines, Paris, 212 pp.

Merriam, D.F. and Jewett, D.G., 1988, Methods of thematic map comparison, in Merriam, D.F., (ed.), Current Trends in Geomathematics: Plenum Press, New York/London, p. 9–17.

Merriam, D.F. and Sneath, P.H.A., 1966, Qualitative comparison of contour maps: Jour. Geophysical Research, v. 71, no. 4, p. 1105–1115.

Merriam, D.F. and Sondergard, M.A., 1988, A reliability index for the pairwise comparison of thematic maps: Geol. Jb., v. A 104, p. 433–446.

Neuman, S.P. and Jacobson, E.A., 1984, Analysis of nonintrinsic spatial variability by residual kriging with application to regional groundwater levels: Math. Geol., v. 16, no. 5, p. 499–521.

Rodionov, D.A., 1981, Statisticěskie resěnija v geologii.: Izd. "Nedra", Moskva, 231 pp.

Tatsuoka, M.M., 1971, Multivariate Analysis—Techniques for Educational and Psychological Research: John Wiley & Sons, New York, 310 pp.

23

THE KINEMATICS OF PALEO LANDFORMS

R.G. Craig

Reconstructions of past climates and other applications of global models require specification of landforms and geomorphic systems as boundary conditions. As general circulation models become more sophisticated and comprehensive and the range of applications for reconstructions grows, there will be an increasing demand for valid geomorphic boundary conditions throughout the Phanerozoic. Geomorphologists have not yet developed the tools and expertise needed to produce reconstructions, so a major gap in understanding of global change now exists. A strategy to fill that gap is presented here.

Challenges and Opportunities

If geomorphology is the study of the form of the land, and if the form of the land can be described by a set of numbers (digital elevation models), then whither geomorphology? Do we become numericists, mathematicians, and statisticians? If geomorphology is the study of the processes shaping the land, and if the form is completely explained by a set of processes and an initial condition (*i.e.*, landform at some earlier time), then two "knowns," current form and current processes, are essential. But if processes themselves change through time and there is an infinite set of initial conditions, one set for each point in time, then the job becomes overwhelmingly complex.

As the geomorphic community has become painfully aware of the difficulties of deep reconstructions, we have withdrawn into the Quaternary, a period during which many simplifying assumptions can be made which allow solution of geomorphic problems. Hence the Geological Society of America lumps "Quaternary Geology" and "Geomorphology" into one division. We have become so comfortable with the notion that geomorphology and

Quaternary geology are synonymous that we have lost sight of the goals of founders of the science such as William Morris Davis and John Wesley Powell who strove to unearth the landforms of the distant past. Of course there is plenty to keep us busy in the good ol' Quaternary; but we shouldn't ignore the enormous challenges and opportunities that await those who would reconstruct landforms of earlier times.

Just as plate tectonics reconstructions have now provided us with "skeleton" representations of the relative positions of the continents at regular intervals throughout the Phanerozoic (Zeigler *et al.*, 1983), a goal of geomorphology should be to provide the "flesh" on such skeletons. As Bloom (1991) stated, "the conceptual goal is a motion picture of a single landscape evolving." We should be able to take on the task of showing the position and character of rivers, mountains, valleys, and plains—the geomorphology on a continent-wide scale—at each time interval. The expansion of knowledge this would represent and the potential applications in fields as diverse as paleoceanography, petroleum geology, stratigraphy, paleoclimatology, and paleontology would be enormous.

Continent-wide geomorphic syntheses can provide a subset of geomorphic variables that is useful to other disciplines. Of great importance to paleoceanographers is the character of sediment being supplied from the continents, its supply point, volume, size distribution, and composition (Beaumont *et al.*, 1988). The dissolved load of rivers is at least as important (Broecker and Peng, 1982; Hodell *et al.*, 1990). With this information, the impact of oceanic currents in redistributing that sediment can be evaluated, sedimentation rates can be estimated, and perhaps the chemical balance of the oceans can be specified (Berner and Berner, 1987). Petroleum geologists may focus upon nearshore facies and the supply of organic nutrients (Jasper and Gagosian, 1989). Paleontologists can use information about the supply of sediments and nutrients. In addition, the temperature of water supplied by the rivers of a continent may be useful (Oglesby *et al.*, 1989; Maasch and Oglesby, 1990). Stratigraphers may concentrate on the clastic load and the variability in supply may be especially important (Chamley, 1979; Flemings and Jordan, 1989; Müller and Spielhagen, 1990). Finally, paleoclimatologists need specific boundary conditions to run their models (Washington and Parkinson, 1986); one of the most important of these is the topographic configuration at a particular time period such as the Late Cretaceous. Also required is the albedo, primarily a function of land cover.

Geomorphic Basis of Continental Reconstructions

Continent-wide reconstruction will require data on the locations of convergent plates, stress fields, and motion vectors. It will require geologic maps in three dimensions to evaluate the evolution of structures supporting the current topography, and knowledge about lithologic units and crustal thickness. We will need climatic estimates for the periods to be reconstructed, paleontologic information to represent land biota controlling exposure, weathering, runoff, soil formation, *etc.*, including the vertical zonation. Finally, we will need specially formulated geomorphic rules of how the landform responds to these various stimuli.

The first great geomorphic synthesizer, William Morris Davis, identified the variables of interest in geomorphic reconstruction. Davis (1889) described three controls on landform evolution, which he called "structure," "process," and "stage." Structure meant all those elements of geology which influenced the landform: geometry of lithologic contacts (folds, faults, *etc.*), character of the rocks, thickness of units, *etc.* Because of the limited accumulation of field observation at the time, "process" had a circumscribed meaning based on conceptual rules defining the behavior of landforms under specified conditions. Under this heading would go Davis' notions of "youth," "maturity," and "old age" and his assumption that streams first adjust to structure at their mouths with the adjustment progressing upstream toward the headwaters. "Stage" was the timekeeper for application of the rules of aging of the landform.

A modern synthesis of the kinetics of a landform must also recognize the importance of "structure," *sensu lato.* Our definition of structure must be adapted to the resolution of the reconstructions attempted, so we may not be able to use directly the precise lithologic character or measured thickness of each unit. If a resolution of 200 km in a reconstruction is a reasonable goal, then "structure" takes on a new meaning and only a statistical summary of features within each cell is usable. For example, we might specify the number of distinct map units and lithologies in a cell, the mean thickness of those units, the average and dispersion of dip, and perhaps transition probabilities for different rock types (*e.g.*, Craig, 1980). Only in an unusually homogeneous region will it be feasible to describe the area in terms of a stratigraphic section and strength of individual rock units. Other data needed for syntheses include the size distribution of structures, degree of lithologic contrast within sections, structural relief, likelihood of having metamorphic or plutonic rocks at the surface, and the lateral continuity of structures. It is from such data that we must infer the geomorphic evolution of a continent.

In the same way, "process" takes on a different meaning in kinematic geomorphology. Modern process studies provide a vast literature on inferential rules (Ritter, 1986), some of which can be applied to continental scale computation. For example, the potential energy available for work can be estimated from the mean elevation in a cell. The variance of potential energy can be based on the variance of elevation. Sufficient data may be obtained to characterize the average cohesion of soils or the shear strength of rocks. We can apply continuity equations to the volume of material removed in each cell, balancing that with sediment delivery to the oceans. But measures of the instability of rock masses within cells can only be defined in a statistical sense, so new conventions for applying process ideas to determine the rate of sediment supply and transport must be specified, perhaps relying on empirical formulae (for example, Flacke *et al.*, 1990; Deploey *et al.*, 1991).

"Stage" retains little of its Davisian meaning. Rather than implying the form of the land when acted upon by specific processes over a specific interval, stage now must be a marker for time. The importance of stage is subsumed by recognition that landform has a Markovian character. Form and structure at a given time is a function of form and structure at the last time step as acted upon by an ensemble of processes (Craig, 1989). Davis, focused on a smaller spatial scale, allowed form to change but did not appreciate the fundamental role of changing structure, as in crustal dynamics (Judson, 1985).

Variables that we should be able to specify in continental reconstruction include elevation, elevation range, and slopes; river patterns including discharge points; discharge of rivers at seasonal resolution; form of hydrograph for typical flows and storms; seasonal distribution of river water temperature; suspended sediment load (composition, quantity, sizes); dissolved sediment load (major ions, concentration); presence of glaciers, deserts, and karst; and land cover types (and related albedoalbedo). We cannot systematically assess anything below the scale of reconstruction, which includes many things that intrigue modern process geomorphologists. This, perhaps, indicates why such a synthesis has yet to be attempted.

A Mechanism for Continental Modeling

First-order controls on landform development include tectonics, structure, lithology, climate, and time. Landform is not a geomorphic response to static tectonic conditions imposed instantaneously at some—usually far removed—past time. Relief exists because the Earth's engine is creating it. To understand the sequential evolution of landforms, we must be able to specify the kinematics of the lithosphere.

For example, volcanism is actively creating topographic relief in the Cascade Range today because this region is on the leading edge of the North American plate and adjacent to the Cascadia subduction zone of the Juan de Fuca and Gorda plates (*cf.* McBirney *et al.*, 1974). Some elements of the landform can be inferred in considerable detail using accepted rules once we know the tectonic setting (McKenzie, 1978; Jordan *et al.*, 1983; Le Pinchon *et al.*, 1988). The Cascades, like other continental margin subduction zone terranes, are expected to be linear and continuous. Relief can be predicted to a first-order approximation from the rate (30 to 40 mm/yr; Heaton and

Kanamori, 1984) and duration of subduction, dip of the subduction zone (10.5° to 15.5°), isostatic considerations, and rock strength (Nur and Ben-Avraham, 1989).

Our understanding of tectonic processes now allows us to predict surface uplift and exhumation rates (England and Molnar, 1990) such that the Cascades will reach an elevation where orographic processes ensure perennial snowpack, glaciers, and attendant alpine topography (Baker *et al.*, 1987). Rivers on the west will drain directly to the Pacific; those on the east will enter an arid region because of the rainshadow effect. These orographic effects are predictable from the latitude and lack of barriers to the west (Whitehouse, 1988; Bannon and Yuhas, 1990; Freytag, 1990). The Cascades are in the belt of westerlies, so orographic lifting will be nearly perpendicular to the mountains and adiabatic cooling very efficient (Barry, 1981). We can predict the elevation at which moist adiabatic processes will characterize the climate and so we can predict the high rainfall and runoff on the west side of the Cascades (Collins and Taylor, 1990; Gurnell, 1990; Shentzis, 1990). Moreover, our understanding of weathering processes on andesitic rocks (themselves predicted by the plate tectonic setting), coupled with the representation of climate, leads to conclusions about soils to be expected (Van Houton, 1982; Tourtelot, 1983; Colman and Dethier, 1986; Yaalon, 1990). We can then infer the character of sediment delivered to the streams (Ahnert, 1970; Harris and Boardman, 1990). The length of stream transport and amount of runoff constrain the sorting, shape, and composition of sediments delivered to the marine environment (Starkel, 1990). Knowledge about climate, soils, and slopes provides guidance on the type of vegetation that should be encountered (Wallace and Oliver, 1990). Avissar and Verstraete (1990) and Henderson-Sellers (1990) review current ways of specifying such factors at this scale and Prentice (1990) focuses on the vegetation component.

The conclusions of modeling will be a guide to *probable* geomorphic character—and the more geomorphic knowledge incorporated into rules defining the system, the more satisfying the result is likely to be.

Geomorphic Synthesis—A Modest Proposal

An ideal system would compute the *dynamics* of the lithosphere/asthenosphere/core in sufficient detail to estimate the outline, state of tectonic stress, structural deformation, and velocity vectors of the continents at any desired time. Initial conditions would be those currently observed and inverse calculations would allow solutions at any time step desired—perhaps every one million years. The information computed would include the uneroded, isostatically unadjusted topographic field. This, in turn, would form the fundamental boundary condition for a fully coupled calculation of the contemporaneous ocean–atmosphere–cryosphere–isostasy system (we are now seeing the first incomplete examples of such calculations at a coarse scale; *e.g.*, Stockmal *et al.*, 1986; Berger *et al.*, 1990; Peltier, 1990; Tushingham and Peltier, 1991).

Oceanic circulation must be specified because it couples with the atmospheric component of the climate system (Hense *et al.*, 1990; Oglesby, 1990; Oglesby and Saltzman, 1990; Wallace *et al.*, 1990) and it defines the offshore record of subaerial geomorphic processes (Shackleton *et al.*, 1990). Oceanic circulation models demonstrate that useful reconstructions of major current systems are possible if the requisite boundary conditions (atmospheric forcing, insolation, and basin geometry) can be specified (Meehl, 1990).

These data would be sufficient to compute the resulting geomorphic systems, including erosion and isostatic adjustment (Pinet and Souriau, 1988). This demonstrates the desirability for a *coupled* ocean–atmosphere–cryosphere–lithosphere–geomorphic calculation (Saltzman, 1990), but such a synthesis is beyond our present capabilities, in part because of the difficulties of tectonic modeling.

Today the geologic community must be satisfied with a sequence of static views of plate (and continental) configurations that are syntheses of field evidence and not computed from first principles. These static views, and the animation they allow, provide a wealth of increased understanding of the geologic history of the Earth. Applications include calculation of scenarios representing Mesozoic climates (Hay *et al.*, 1982; 1990), experimentation with the factors responsible for Carboniferous glaciations (Beran *et al.*, 1991), unraveling paleontologic puzzles (Parrish *et al.*, 1986), reconstructing paleoceanographic circulation of the Mesozoic and Tertiary (Barron and Peterson, 1989; Rind and Chandler, 1991), and improving the exploration for oil (Kruijs and Barron, 1990). Can we achieve the same benefits from geomorphic synthesis?

A Realistic Proposal for Kinematic Reconstructions

Canonical reconstructions of geomorphic systems achieve surprising detail. In the case of the modern Cascade Mountains, all information can be inferred from available plate tectonic reconstructions and geomorphic systems. Difficulties increase with increasing age of the reconstruction, as is expected (Richards and Engebretson, 1992). For example, inference about stresses in plates depends on assumed rates of motion (Engebretson *et al.*, 1985), themselves uncertain because of inaccuracies in dating. Evolution of plant species limits understanding of their association with climate. Climatic reconstructions are limited by our knowledge of critical variables such as ocean paleosalinity, atmospheric composition, and solar insolation (Stone and Risbey, 1990). We can expect greater accuracy for more recent times, so it makes sense to begin with the sequence of the Cenozoic.

Indeed, reconstruction should start with the modern landform. A set of rules to "reconstruct" it must first be developed. One starting point is the Cooperative Holocene Mapping Project (COHMAP, 1988) for specifying the environmental setting during the last major glaciation. We could next concentrate on the Cenozoic at 10-million-year increments. Plate tectonics and other geologic history are sufficiently well known to allow the effort and to constrain its path (Ziegler *et al.*, 1983). Only after the Cenozoic has been reconstructed would it make sense to attempt to reconstruct the Mesozoic. At these time scales, palinspastic reconstructions become more important and 10-million-year time steps are about as ambitious as is practical. Finally, the Paleozoic could be attempted; reconstruction would be less constrained, but availability of rule-based reconstructions would be immensely beneficial to stratigraphers and paleoceanographers. We can assume that solutions are independent between continents, so each can be reconstructed separately. Our work can begin with one continent, and one era—the Cenozoic.

Horizontal scales on the order of several hundreds of kilometers can be expected. Existing GCM paleoclimate reconstructions are available at scales such as $4.5° × 7.5°$ (Washington and Parkinson, 1986). Experimental studies of modern climate rarely exceed $2.25° × 3.75°$ (Gordon and Stern, 1982). Mesoscale models covering the western one-third of the continent commonly operate at 60-km spacing (Giorgi, 1990), and it is usually recommended that GCM results should be applied using averages of a $3 × 3$ neighborhood cell (Grotch and MacCracken, 1991). Therefore, a 200-km grid for geomorphic reconstructions seems optimistic in its assumption of data availability, and even this requires about 5000 points to represent North America. At a resolution of 200 km, inferences about the geomorphic system can be made for elevation, elevation range, and slopes; drainage direction and discharge points; climate (seasonal temperature, precipitation, and evaporation); seasonal presence of snow or ice; major land-forming processes at work; soil types and thickness; dominant vegetation types; erosion and denudation rates; relative magnitudes of chemical and physical weathering. We can imagine at least 30 variables specified for each time step, giving a total of about a million items. Answers cannot be computed in an analytic sense, because we lack a complete specification of the appropriate equations.

An alternative is to infer answers using the type of rules previously applied to the Cascade Mountains. These rules must depend on more than continental configuration as boundary conditions. Not enough is known about plate tectonics to predict uplift of the western cordillera at a scale of 200 km and 10 million years. Additional constraints are needed in the form of empirical evidence from the geologic record (*e.g.*, Hays and Grossman, 1991). For example, the Eocene topography of northwestern Wyoming can be inferred because fossils of the Willwood Formation imply an altitude of less than 2000 m (Wing and Bown, 1985) and a warm, equable climate (Wing *et al.*, 1991). Combining this with similar geologic records of adjacent points imposes significant constraints. For example, glaciers would be unlikely, sediment supply abundant, and the hydrograph moderately flat (Collins and Taylor, 1990).

If a variable is fixed by the geologic record, some assignment of confidence should be made that can be propagated through subsequent inferences. The inference engine must be flexible enough to incorporate this uncertainty. Simple rules for determining if glaciers exist in the cell under consideration might include:

```
FOR ( SEASON = WINTER; SEASON <= FALL; SEASON++){
    IF ( TEMP [SEASON] < 0 ) THEN ACCUMULATION += PPT [SEASON];
    IF ( TEMP [SEASON] > 0 ) THEN ABLATION += EVAPORATION [SEASON];
    ICE_BALANCE = ICE_BALANCE + ACCUMULATION - ABLATION;}
```

This assumes that four temperatures and precipitations are available, perhaps specified as a solution of a GCM for this continental configuration (Washington and Parkinson, 1986). Otherwise they must be inferred from the geologic record. For example, elevation and latitude can be estimated from the plate tectonic setting (Whitehouse, 1988). We can apply a lapse rate to the elevation to estimate temperature (Barry, 1981); but we must also know temperature at sea level, and sea level itself (Haq *et al.*, 1987). Sea-level temperature could be estimated from a latitude-dependent energy balance (Jentsch, 1991); but adjustment for moist adiabatic cooling upwind may be needed, in which case winds must be known (Ersi, 1991). If GCM results do not provide wind estimates, we may use geologic evidence (Lancaster, 1984) or general meteorologic reasoning (Bannon and Yuhas, 1990; Emeis, 1990).

Calibrating rules is not merely the searching of the literature for clues to the status of the Earth at specific times. Some of the rules must be made up (*cf.* Mayer, 1985). For instance, elevation is greater in the west than in the east, presumably because the west is closer to the leading edge of a plate. A naive rule is "the closer a point to the leading edge of a plate, the greater its altitude." This leads to triangular cross-sections, which are very crude approximations.

The rules may be refined to yield better inferences. Elevation is the fundamental geomorphic variable and tectonic activity its most important control, so we begin with the relationship between the two. A better understanding may be achieved through analysis of a database of elevations and digital line graphs of plate tectonic elements. It is possible to determine the distances between every point and every plate tectonic element in the database. A statistical model of the relation between plate tectonic geometry and elevations can be built for the smaller distances. Variables of interest might include direction of movement relative to the long axis of the plate tectonic element and speed with respect to that element. If successful, statistical rules would be available to generate the landform, in the first step toward synthesizing the kinematics of landforms.

Caveats

The uncertainty involved in such global inferences is very large and may dominate the solution. The complexity of rules required is daunting and may be overwhelming. The magnitude of the task of gathering and interpreting the geologic record is frightful. The entire effort is enormous even within the limits imposed by our lack of understanding of the physics of the Earth's system (Saltzman, 1990). However, the potential benefits of such a synthesis to our understanding of the progression of terrestrial landforms through time are enormous.

Acknowledgments

I thank Jon Harbor and Dave Smith for reviews of earlier drafts of the manuscript. I remember Dotty Craig for her special help; I couldn't have done it without her.

References

Ahnert, F., 1970, Functional relationships between denudation, relief, and uplift in large mid-latitude drainage basins: *Am. Jour. Sci.*, v. 68, p. 243–263.

Avissar, R. and Verstraete, M.M., 1990, The representation of continental surface processes in atmospheric models: *Reviews of Geophysics*, v. 28, no. 1, p. 35–52.

Baker, V.R., Greeley, R., Komar, P.D., Swanson, D.A. and Waitt, R.B., Jr., 1987, Columbia and Snake River plains, *in* Graf, W.L., (ed.), *Geomorphic Systems of North America, The Decade of North American Geology:* Centennial Spec. Vol., Geol. Soc. Amer., v.2, p. 403–468.

Bannon, P.R. and Yuhas, J.A., 1990, On mountain wave drag over complex terrain: *Meteorology and Atmospheric Physics*, v. 43, no. 1–4, p. 155–162.

Barron, E.J. and Peterson, W.H., 1989, Model simulations of the Cretaceous ocean circulation: *Science*, v. 244, p. 684–686.

Barry, R.G., 1981, *Mountain Weather and Climate:* Methuen, New York, 313 pp.

Beaumont, C., Quinlan, G.M. and Hamilton, J., 1988, Orogeny and stratigraphy: Numerical models of the Paleozoic in eastern North America: *Tectonics*, v. 7, p. 389–416.

Beran, M.A., Brilly, M., Becker, M., Bonacci O., Crowley, T.J., Baum, S.K. and Hyde, W.T., 1991, Climate model comparison of Gondwanan and Laurentide glaciations: *Jour. Geophysical Research–Atmospheres*, v. 96, no. D5, p. 9217–9226.

Berger, A., Gallee, H., Fichefet, T., Marsiat, I. and Tricot, C., 1990, Testing the astronomical theory with a coupled climate ice-sheet model: *Global and Planetary Change*, v. 89, no. 1–2, p. 125–141.

Berner, E.A. and Berner, R., 1987, *The Global Water Cycle Geochemistry and Environment:* Prentice–Hall, Englewood Cliffs, N.J., 397 pp.

Bloom, A.L., 1991, *Geomorphology: A Systematic Analysis of Late Cenozoic Landforms, 2nd Ed.:* Prentice–Hall, Englewood Cliffs, N.J., 532 pp.

Broecker, W.S. and Peng, T.-H., 1982, *Tracers in the Sea:* Eldigio Press, Palisades, New York, 634 pp.

Chamley, H., 1979, North Atlantic clay sedimentation and paleoenvironment since the Late Jurassic: Deep drilling results in the Atlantic Ocean, *in* Talwani, M., Hay, W. and Ryan W.B.F., (eds.), *Continental Margins and Paleoenvironment*, Ewing Series, Vol. 3: Am. Geophys. Union, Washington, D.C., p. 342–361.

COHMAP, 1988, Climatic changes of the last 18,000 years: Observations and model simulations: *Science*, v. 241, p. 1043–1052.

Collins, D.N. and Taylor, D.P., 1990, Variability of runoff from partially glacierised alpine basins: *Hydrology in Mountainous Regions I*, IAHS Publications 193, p. 365–372.

Colman, S.M. and Dethier, D.P., 1986, An overview of rates of weathering, *in* Colman, S.M. and Dethier, D.P., (eds.), *Rates of Chemical Weathering of Rocks and Minerals*: Acad. Press, Orlando, Florida, p. 1–18.

Craig, R.G., 1980, A computer program for the simulation of landform erosion: *Computers & Geosci.*, v. 6, p. 111–142.

Craig, R.G., 1989, Computing Appalachian geomorphology: *Geomorphology*, v. 2, p. 197–207.

Davis, W.M., 1889, The rivers and valleys of Pennsylvania: *National Geographic Magazine*, v. 1, p. 183–253.

Deploey, J., Kirkby, M.J. and Ahnert F., 1991, Hillslope erosion by rainstorms—A magnitude frequency-analysis: *Earth Surface Processes and Landforms*, v. 16, no. 5, p. 399–409.

Emeis, S., 1990, Surface pressure distribution and pressure drag on mountains: *Meteorology and Atmospheric Physics*, v. 43, no. 1–4, p. 173–185.

Engebretson, D.C., Cox, A. and Gordon, R.G., 1985, Relative motions between oceanic and continental plates in the Pacific Basin: *Geol. Soc. Amer. Special Paper*, v. 206, p. 59.

England, P. and Molnar, P., 1990, Surface uplift, uplift of rocks and exhumation rates: *Geology*, v. 18, p. 1173–1177.

Ersi K., 1991, Relationship between runoff and meteorological factors and its simulation in a Tianshan glacierized basin, *in* Bergmann, H., *et al.*, (eds.), *Snow, Hydrology and Forests in High Alpine Areas*: IAHS Publications 205, p. 189–202.

Flacke, W., Auerswald, K. and Neufang, L., 1990, Combining a modified universal soil loss equation with a digital terrain model for computing high-resolution maps of soil loss resulting from rain wash: *Catena*, v. 17, no. 4-5, p. 383–397.

Flemings, P.B. and Jordan, T., 1989, A synthetic stratigraphic model of foreland basin development: *Jour. Geophysical Research*, v. 94, B4, p. 3851–3866.

Freytag, C., 1990, Modifications of the structure of cold fronts over the foreland and in a mountain valley: *Meteorology and Atmospheric Physics*, v. 43, no. 1–4, p. 69–76.

Giorgi, F., 1990, Simulation of regional climate using a limited area model nested in a general-circulation model: *Jour. of Climate*, v. 3, no. 9, p. 941–963.

Gordon, C.T. and Stern, W., 1982, A description of the GFDL global spectral model: *Monthly Weather Review*, v. 110, no. 7, p. 625–644.

Grotch, S.L. and MacCracken, M.C., 1991, The use of general-circulation models to predict regional climatic change: *Jour. of Climate*, v. 4, no. 3, p. 286–303.

Gurnell, A.M., 1990, Improved methods of assessment of snow and glaciers as water-balance and river flow components: *Hydrology in Mountainous Regions I*, IAHS Publications 193, p. 157–172.

Haq, B.U., Hardenbol, J. and Vail, P.R., 1987, Chronology of fluctuating sea levels since the Triassic: *Science*, v. 235, p. 1156–1167.

Harris, T. and Boardman J., 1990, A rule-based expert system approach to predicting waterborne soil-erosion: *Soil Erosion on Agricultural Land*, British Geomorphological Research Group Symposia Series, p. 401–412.

Hay, W.W., Barron, E.J. and Thompson, S.L., 1990, Global atmospheric circulation experiments on an earth with polar and tropical continents: *Jour. Geol. Soc.*, v. 147, p. 749–757.

Hay, W.W., Behensky, J.F., Jr., Barron, E.J. and Sloan, J.L., 1982, Late Triassic-Liassic paleoclimatology of the proto-central north Atlantic rift system: *Palaeogeography, Palaeoclimatology, Palaeoecology*, v. 40, p. 13–30.

Hays, P.D. and Grossman, E.L., 1991, Oxygen isotopes in meteoric calcite cements as indicators of continental paleoclimate: *Geology*, v. 19, no. 5, p. 441–444.

Heaton, T.H. and Kanamori, H., 1984, Seismic potential associated with subduction in the northwestern United States: *Seismol. Soc. Amer. Bull.*, v. 74, p. 933–941.

Henderson-Sellers, A., 1990, The coming of age of land surface climatology: *Global and Planetary Change,* v. 82, no. 3–4, p. 291–319.

Hense, A., Glowienkahense, R., Vonstorch, H. and Stahler, U., 1990, Northern-hemisphere atmospheric response to changes of Atlantic-ocean SST on decadal time scales—A GCM experiment: *Climate Dynamics,* v. 4, no. 3, p. 157–174.

Hodell, D.A., Mead, G.A. and Mueller, P.A., 1990, Variation in the strontium isotopic composition of seawater (8 ma to present)—Implications for chemical-weathering rates and dissolved fluxes to the oceans: *Chemical Geology,* v. 80, no. 4, p. 291–307.

Jasper, J.P. and Gagosian, R.B., 1989, Alkenone molecular stratigraphy in an oceanic environment affected by glacial freshwater events: *Paleoceanography,* v. 4, p. 603–614.

Jentsch, V., 1991, An energy-balance climate model with hydrological cycle. 1. Model description and sensitivity to internal parameters: *Jour. Geophysical Research–Atmospheres,* v. 96, no. D9, p. 17169–17179.

Jordan, T.E., Isacks, B.L., Ramos, V.A. and Allmendinger, R.W., 1983, Mountain building in the central Andes: *Episodes,* v. 1983, no. 3, p. 20–26.

Judson, S., 1985, Some observations on paleogeomorphology, *in* Hayden, R.S., (ed.), *Global Mega-Geomorphology:* NASA Conference Publication 2312, p. 27–28.

Kruijs, E. and Barron, E., 1990, Climate model prediction of paleoproductivity and potential source-rock distribution: Deposition of organic facies: Am. Assoc. Petroleum Geologists *Studies in Geology,* v. 30, p. 195–216.

Lancaster, N., 1984, Characteristics and occurrence of wind erosion features in the Namib Desert: *Earth Surface Processes,* v. 9, no. 5, p. 469–478.

Le Pinchon, X., Bergerat, F. and Roulet, M.-J., 1988, Plate kinematics and tectonics leading to the Alpine belt formation; a new analysis, *in* Clark, S.P., Jr., Burchfiel, B.C. and Suppe, J., (eds.), *Processes in Continental Lithosphere Deformation:* Geol. Soc. Amer. Spec. Paper 218, p. 111–131.

Maasch, K.A. and Oglesby, R.J., 1990, Meltwater cooling of the Gulf of Mexico: A GCM simulation of climatic conditions at 12 ka: *Paleoceanography,* v. 5, no. 6, p. 977–996.

Mayer, L., 1985, Quantitative analysis in megageomorphology, *in* Hayden, R.S., (ed.), *Global Mega-Geomorphology:* NASA Conference Publication 2312, p. 79–80.

McBirney, A.R., Sutter, J.F., Naslund, H.R., Sutton, K.G. and White, C.M., 1974, Episodic volcanism in the central Oregon Cascade Range: *Geology,* v. 2, p. 585–589.

McKenzie, D., 1978, Some remarks on the development of sedimentary basins: *Earth and Planetary Sci. Letters,* v. 40, p. 25–32.

Meehl, G.A., 1990, Development of global coupled ocean-atmosphere general-circulation models: *Climate Dynamics,* v. 5, no. 1, p. 19–33.

Müller, R.D. and Spielhagen, R.F., 1990, Evolution of the central Tertiary basin of Spitsbergen—Towards a synthesis of sediment and plate tectonic history: *Palaeogeography, Palaeoclimatology, Palaeoecology,* v. 80, no. 2, p. 153–172.

Nur, A. and Ben-Avraham, Z., 1989, Oceanic plateaus and the Pacific Ocean margins, *in* Ben-Avraham, Z., (ed.), *The Evolution of the Pacific Ocean Margins:* Oxford Univ. Press, New York, p. 7–19.

Oglesby, R.J., 1990, Sensitivity of glaciation to initial snow cover, CO_2, snow albedo, and oceanic roughness in the NCAR CCM: *Climate Dynamics,* v. 4, no. 4, p. 219–235.

Oglesby, R.J., Maasch, K.A. and Saltzman, B., 1989, Glacial meltwater cooling of the Gulf of Mexico: GCM implications for Holocene and present-day climate: *Climate Dynamics,* v. 3, no. 3, p. 115–133.

Oglesby, R.J. and Saltzman, B., 1990, Extending the EBM—The effect of deep ocean temperature on climate with applications to the Cretaceous: *Global and Planetary Change,* v. 82, no. 3–4, p. 237–259.

Parrish, J.M., Parrish, J.T. and Zeigler, A.M., 1986, Permian-Triassic paleogeography and paleoclimatology and implications for therapsid distribution, *in* Holton, N., III *et al.,* (eds.), *The Ecology and Biology of Mammal-like Reptiles:* Smithsonian Institution Press, Washington, D.C., p. 109–145.

Peltier, W.R., 1990, Paleoenvironmental modeling and global change: *Global and Planetary Change,* v. 82, no. 1–2, p. 79–85.

Pinet, P. and Souriau, M., 1988, Continental erosion and large-scale relief: *Tectonics,* v. 7, p. 563–582.

Prentice, K.C., 1990, Bioclimatic distribution of vegetation for general-circulation model studies: *Jour. Geophysical Research–Atmospheres*, v. 95, no. D8, p. 11811–11830.

Richards, M.A. and Engebretson, D.C., 1992, Large-scale mantle convection and the history of subduction: *Nature*, v. 355, p. 437–440.

Rind, D. and Chandler, M., 1991, Increased ocean heat transports and warmer climate: *Jour. Geophysical Research–Atmospheres*, v. 96, no. D4, p. 7437–7461.

Ritter, D.F., 1986, *Process Geomorphology, 2nd Ed.*: W.C. Brown Co., Dubuque, Iowa, 579 pp.

Saltzman, B., 1990, Three basic problems of paleoclimatic modeling—A personal perspective and review: *Climate Dynamics*, v. 5, no. 2, p. 67–78.

Shackleton, N.J., Van Andel, T.H., Boyle, E.A., Jansen, E., Labeyrie, L., Leinen, M., McKenzie, J., Mayer, L. and Sundquist, E., 1990, Contributions from the oceanic record to the study of global change on three time scales—Report of Working Group 1: Interlaken Workshop for Past Global Changes, *Global and Planetary Change*, v. 82, no. 1-2, p. 5–37.

Shentzis, I.D., 1990, Mathematical models for long-term prediction of mountainous river runoff—Methods, information and results: *Hydrological Sci. Jour.–Jour. des Sciences Hydrologiques*, v. 35, no. 5, p. 487–500.

Starkel, L., 1990, Global continental paleohydrology: *Global and Planetary Change*, v. 82, no. 1-2, p. 73–77.

Stockmal, G.S., Beaumont, C. and Boultier, R., 1986, Geodynamic models of convergent margin tectonics: Transition from rifted margin to overthrust belt and consequences for foreland-basin development: *Bulletin Am. Assoc. Petroleum Geologists*, v. 70, p. 181–190.

Stone, P.H. and Risbey, J.S., 1990, On the limitations of general-circulation climate models: *Geophys. Research Letters*, v. 17, no. 12, p. 2173–2176.

Tourtelot, H.A., 1983, Continental aluminous weathering sequences and their climatic implications in the United States, *in* Cronin, T.M., Cannon, W.F. and Poore, R.Z., (eds.), *Paleoclimate and Mineral Deposits*: U.S. Geological Survey, Circ. 822, p. 1–5.

Tushingham, A.M. and Peltier, W.R., 1991, Ice-3G—A new global-model of Late Pleistocene deglaciation based upon geophysical predictions of postglacial relative sea-level change: *Jour. Geophysical Research–Solid Earth and Planets*, v. 96, no. B3, p. 4497–4523.

Van Houten, F.B., 1982, Ancient soils and ancient climates: *Climate in Earth History*, National Academy Press, Washington, D.C., p. 112–117.

Wallace, J.S. and Oliver, H.R., 1990, Vegetation and hydroclimate, *in* Anderson, M.G. and Burt, T.P., eds., *Process Studies in Hillslope Hydrology*: John Wiley & Sons, New York, p. 9–41.

Wallace, J.M., Smith, C. and Jiang, Q.R., 1990, Spatial patterns of atmosphere ocean interaction in the northern winter: *Jour. of Climate*, v. 3, no. 9, p. 990–998.

Washington, W.M. and Parkinson, C.L., 1986, *An Introduction to Three-Dimensional Climate Modeling*: University Science Books, Mill Valley, Calif., 422 pp.

Whitehouse, I.E., 1988, Geomorphology of the central Southern Alps, New Zealand: The interaction of plate collision and atmospheric circulation: *Zeitschrift für Geomorph.*, Supp. Bd., v. 69, p. 105–116.

Wing, S.L. and Bown, T.M., 1985, Fine-scale reconstruction of late Paleocene–early Eocene paleogeography in the Bighorn basin of northern Wyoming, *in* Flores, R.M. and Kaplan, S.S., (eds.), *Cenozoic Paleogeography of West-Central United States*: Soc. Econ. Paleontologists and Mineralogists, Rocky Mountain Section, p. 93–105.

Wing, S.L., Bown, T.M. and Obradovich, J.D., 1991, Early Eocene biota and climatic changes in interior western North America: *Geology*, v. 19, p. 1189–1192.

Yaalon, D., 1990, The relevance of soils and paleosols in interpreting past and ongoing climatic changes: *Global and Planetary Change*, v. 82, no. 1–2, p. 63–64.

Ziegler, A.M., Scotese, C.R. and Barrett, S.F., 1983, Mesozoic and Cenozoic paleogeographic maps, *in* Brosche, P. and Sunderman, J., (eds.), *Tidal Friction and the Earth's Rotation, II*: Springer–Verlag, Berlin, p. 240–252.

24

R.G.V. EIGEN: LEGENDARY FATHER OF MATHEMATICAL GEOLOGY

J.H. Doveton and J.C. Davis

Prologue

The original 1968 constitution of the International Association for Mathematical Geology provided that a newsletter be distributed to members. Initially the newsletter drew limited attention and consisted of little more than membership lists and postings of proposed revisions to the by-laws. As time went on, the content expanded to news of conferences and topics of interest to the geoscience community. Then a newsletter event sparked extraordinarily wide interest among the membership: the appearance in 1975 of a series on the life and contributions of R.G.V. Eigen.

The idea was born during a luncheon hosted by John Harbaugh at the Eldridge Hotel in Lawrence, Kansas. Because Manfred Eigen was a guest, the conversation naturally turned to the sad neglect of the memory of his illustrious namesake. It was also observed that geomathematicians owed a special debt to R.G.V. Eigen as the earliest author of a mathematical geology paper. Clearly, the *IAMG News Letter* would be an ideal forum to recall the achievements of this great pioneer, as an inspiration to the society members and as a simple matter of justice. The publication of the series resulted in a number of letters to the *News Letter* editor, John Davis. Most of the correspondents threw light on previously unknown incidents in Eigen's life as well as contributing items on Eigen's colleagues and rivals. Other, more cynical readers claimed to find an uncanny resemblance between pictures of Eigen and D.F. Merriam. However, the voices of these skeptics were stilled as anecdotal evidence of Eigen's career poured in from around the world. To ensure that the legacy of this legendary Father of Mathematical Geology will be preserved for future generations of mathematical geologists, much of the currently known Eigen material from the *News Letter* series and ensuing

Figure 1: R.G.V. Eigen (1833–76).

correspondence is summarized here, as well as a previously unpublished account of Eigen's meeting with Lewis Carroll.

Early Life

Rudolf Gottlieb Viktor Eigen (1833–1876) is remembered by all mathematical geologists for his elucidation of the matrix properties that bear his name (Fig. 1). Less well known are his pioneer mathematical investigations of glacial topology that distinguish him as possibly the true "father of mathematical geology." Eigen was born in the Austrian mountain village of Heiligenblut, the second son of a local doctor. Following a brilliant student career at the University of Göttingen, he became the youngest professor of mathematics in the history of that illustrious university. The young professor was highly popular with his students although his stringent academic demands earned him the nickname "Eiger" (ogre). Eigen's bohemian lifestyle resulted in conflict with the university authorities. This led ultimately to his enforced resignation following a scandal that rocked the university and purportedly involved no less than five professors' daughters. The embittered Eigen returned to Heiligenblut and devoted himself to his second love of mountaineering. The mathematician eked out a modest living by acting as mountain guide for rich Victorian climbers and supplemented his income by delivering invited lectures to mathematical societies in England and France.

The Eigen Gambit

Eigen was also a gifted, if highly unorthodox, chess player. His interest in the game stemmed partly from a belief that chess could be approached as an 8 × 8 matrix problem and positional play mathematically analyzed in terms of *Kernvektoren*. Eigen's greatest moment in chess occurred when he met Paul Morphy (considered by some to be the greatest chess player of all time) in Paris in 1858 (Fig. 2). A casual game between the two men at the

Figure 2: Rudolf Eigen draws a chess game with Paul Morphy at the Café Guerbois in 1858.

Café Guerbois resulted in a draw following an aggressive, if bizarre, opening by Eigen:

WHITE: R.G.V. EIGEN   BLACK: P. MORPHY

1.	KT–QB3	P–K4
2.	P–B4	P × P
3.	Kt–Q5!	

The opening (still known as the "Eigen Gambit") anticipated Réti's hypermodern school by some fifty years (Golombek, 1956). Eigen himself invariably referred to his opening as *"die Primzahlen im Reigen"* (dance of the prime numbers). During his brief acquaintance with Morphy, the young New Orleans lawyer introduced Eigen to the mint julep which became one of Eigen's enduring passions and one which he found difficult to satisfy in his native Austria. The Eigen Gambit remains a chess oddity and is rarely, if

Figure 3: "The White Rabbit" drawn by Sir John Tenniel (Carroll, 1865).

ever, used in grandmaster play. However, it acquired a mild notoriety in the chess world when it was selected as a central case study in Sigmund Freud's *Die Psychopathologie des Schachspieles* (Freud, 1912). It would seem that Freud mistook Eigen's matrix approach to chess as symptomatic of an advanced patricidal-anal rejection syndrome.

The White Rabbit

Eigen was acquainted with many of the prominent European mathematicians of the day, including Charles Dodgson ("Lewis Carroll") who may have used him as the inspirational basis for the *Alice* character, "the White Rabbit" (Fig. 3). The sole evidence for this theory is contained in a letter from Carroll to Mrs. Edna Matlock (Warminster, 1932, p. 37–9), as follows:

> "Our convivial picnic group was joined by the Austrian mathematician, Rudolph [*sic*] Eigen and his young French acolyte, François Propère. His continental mien reminds one irresistibly of his homeland, immortalized in the words of Mr. Tennyson,
>
> > *Where ne'er the pluméd Phrygian helm*
> > *Has lofted o'er the samite peaks*   [*1852*]
>
> Dr. Eigen's vivid improvisations and ingenious conjuring tricks were warmly received by the children, and even Alice was moved to laughter in spite of her lingering chill. Dr. Eigen was kind enough to prepare for her a draught of 'magic potion' as he

called it, and I was surprised to see that in next to no time, the colour had returned to her cheeks and she became quite animated. He confided to me that the herbal preparation was of American Indian origin. It has a curious name and he spelled it out for me; it is written thus, 'Mint Julep.' Dr. Eigen fervently maintained that this was the greatest gift of the New World to the Old, surpassing even the Potato and Tobacco in that regard. At length, Dr. Eigen consulted his time-piece, became considerably agitated, and departed in some haste to keep an engagement with his hosts at All Souls. Alice was quite taken with Dr. Eigen and entreated me to tell her when we would next see 'the White Rabbit,' as she mischievously calls him."

Mysterious Death

Eigen published little and is chiefly remembered for his monumental *Überbrückungsschlüsse der Zahlen* (1867; out of print for many years) and his fragmentary and vituperative correspondence with Cayley (originator of matrix algebra). Eigen's life was prematurely cut short when he fell to his death in a crevasse on the Pasterzen glacier following a successful ascent of the Grossglockner. On the centenary of his death, a bronze plaque was unveiled at a bend of the Hochalpenstrasse overlooking the Pasterzen. At that time, the circumstances of Eigen's death were hotly disputed by the eminent Russian mathematical geologist, Andrew B. Vistelius. He asserted that Eigen had survived his fall and had been rescued by the famous writer, Kuz'ma Prutkov. By his account, Eigen accompanied Prutkov back to Russia, embraced the Russian Orthodox faith, and worked as a missionary to the wild "*Chepookha*" tribes in remote parts of Siberia (Fig. 4). Sustained by this intensely spiritual lifestyle, Eigen survived to the grand age of 111.

The Métro Station

It is not generally known that for a long period Rudolf Eigen had a memorial to his genius in the form of a small station on the Paris Métro system (Fig. 5). The first line of the Métro was built between Porte de Vincennes and Porte Maillot and came into operation in 1900. In a surprise decision, the Directoire du Métro designated a station to be named in honor of Eigen in recognition both of his scientific contributions and almost legendary standing with Parisians of his day. However, the choice of an Austrian mathematician had pointed political overtones and was used as part of an abortive diplomatic initiative to drive a wedge between the Austrian and German emperors. Théophile Delcassé, the French minister for foreign affairs, presented the

Figure 4: *Vivit post funera virtus.* R.G.V. Eigen [*in dubio*] with Chepookha tribesman.

station keys to the Emperor Franz Josef in a formal court ceremony at the Schönbrunn Palace in Vienna.

The station would have receded into the anonymity of the rest of the Métro network but for the unusual role it was to play during the Second World War. In 1940, German forces commandeered the station and made it headquarters for the Zugkommando (Zk) Gruppe III, a highly specialized combat unit of Wehrmacht railway engineers (Ernst, 1967). The Zk units had a poorly defined but ambitious role in Operation Sealion (the projected invasion of England). Their function focused on a planned amphibious assault executed from E-boats striking up the Thames and aimed at the paralysis or control of London's railway termini and Underground system.

With the wane of interest in Operation Sealion, control of the station shifted to the Luftwaffe and Eigen station was eventually used as a shipping point for paintings intended for Göring's art collection. The station met its end at the hands of German army demolition squads in the withdrawal from Paris in 1944. Economics of Métro operation mitigated against reopening the station, and the name Eigen has disappeared from post-war Métro maps.

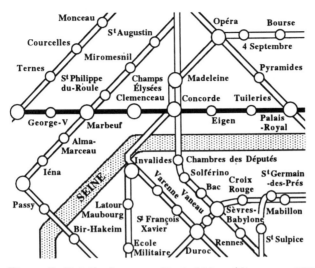

Figure 5: Detail of pre-war Paris Métro (Queneau, 1936).

Eigen Redux

It is astonishing that the name of Eigen might have languished in total obscurity had it not been for the diligent researches of members of the International Association for Mathematical Geology and their dogged determination that once rediscovered, this household name should not slide back into the abyss of ignorance. A request was made to the readership of the *IAMG News Letter* for more information on Eigen in preparation for the centennial commemoration of his death. Early speculations by J.W. Harbaugh were soon followed by correspondence from noted researchers such as W.W. Schwarzacher, R.A. Reyment, and A.B. Vistelius among others, including those who facetiously reported a resemblance between Eigen and a somewhat-celebrated modern and oft-bearded mathematical geologist.

Professor Schwarzacher pointed out that some authors prefer to describe latent roots as "Proper values" rather than "Eigen values." This alternative terminology honors the eminent French mathematician, François Propère, who allegedly discovered the latent roots of matrices almost simultaneously with Eigen. Although the priority of discovery has never been established, we note two observations that tend to favor Eigen. First, anglophone authors are notorious for their cavalier treatment of French words, but it is unusual to honor a mathematician by misspelling his name. Second, in the correspondence quoted earlier, it is clear who Charles Dodgson considered the mentor and who the student when he entertained the two mathematicians.

Professor Reyment discovered a veritable treasure trove of Eigen documents in an ancient briefcase in an attic at the Sthlm Geological Survey.

Figure 6: The late Geoff Hill tracks down evidence of an Eigen visit to the École des Mines de Paris.

Although it will probably take years to decipher all the writings, some of the material suggested that Eigen was interested in linguistics, particularly as it was involved with Gypsy lore. The dedicated eigenist, Geoff Hill, took great pleasure in sharing his findings with his colleagues and discovered definitive evidence for a visit by Eigen to the École des Mines de Paris, at Fontainebleau (Fig. 6).

Finally, as a tribute that would surely have brought Eigen immense pride, the Rudolf G. V. Eigen Colloquium was established at the Institute for Geology of the Free University of Berlin. The annual colloquium was organized by Professor Wolfdietrich Skala, and the first meeting was held in 1978 with six presentations on research topics in mathematical geology.

References

Carroll, L., 1865, *Alice's Adventures in Wonderland:* MacMillan & Co., London, 192 pp.

Eigen, R.G.V., 1867, *Überbrückungsschlüsse der Zahlen:* Ostpresse, München, 894 pp.

Ernst, R., 1967, *Zugkommando!:* Blindt, Berlin, 210 pp.

Freud, S., 1912, Die Psychopathologie des Schachspieles: *Internationale Zeitschrift für ärztliche Psychoanalyse,* v. 4, p. 217-28.

Golombek, H., 1956, The Eigen Gambit: False Dawn of the Hypermodern School: *British Chess Magazine,* v. *xii,* no. 4, p. 326–32.

Queneau, F., 1936, *L'histoire magnifique du Métropolitain:* Éditions Oiseaux, Paris, 118 pp.

Tennyson, A.L., 1852, "Alpine Fastness", *in* Cornwell, J.S., (ed.), 1932, *God's Denunciations against Pharaoh—Hophra and Other Poems:* Macmillan, London, 312 pp.

Warminster, B.C.G., 1932, *Charles Dodgson: The Cherwell Letters:* Macmillan, London, 238 pp.

Index